Nadine Ziemert

Ribosomal biosynthesis of two cyclic peptides from cyanobacteria

AF061197

Nadine Ziemert

Ribosomal biosynthesis of two cyclic peptides from cyanobacteria

Identification and characterisation of microcyclamide and microviridin biosynthesis pathways in freshwater cyanobacteria

Südwestdeutscher Verlag für Hochschulschriften

Impressum/Imprint (nur für Deutschland/ only for Germany)
Bibliografische Information der Deutschen Nationalbibliothek: Die Deutsche Nationalbibliothek verzeichnet diese Publikation in der Deutschen Nationalbibliografie; detaillierte bibliografische Daten sind im Internet über http://dnb.d-nb.de abrufbar.

Alle in diesem Buch genannten Marken und Produktnamen unterliegen warenzeichen-, marken- oder patentrechtlichem Schutz bzw. sind Warenzeichen oder eingetragene Warenzeichen der jeweiligen Inhaber. Die Wiedergabe von Marken, Produktnamen, Gebrauchsnamen, Handelsnamen, Warenbezeichnungen u.s.w. in diesem Werk berechtigt auch ohne besondere Kennzeichnung nicht zu der Annahme, dass solche Namen im Sinne der Warenzeichen- und Markenschutzgesetzgebung als frei zu betrachten wären und daher von jedermann benutzt werden dürften.

Verlag: Südwestdeutscher Verlag für Hochschulschriften Aktiengesellschaft & Co. KG
Dudweiler Landstr. 99, 66123 Saarbrücken, Deutschland
Telefon +49 681 37 20 271-1, Telefax +49 681 37 20 271-0
Email: info@svh-verlag.de
Zugl.: Berlin, Humboldt Universität zu Berlin, Dissertation, 2009

Herstellung in Deutschland:
Schaltungsdienst Lange o.H.G., Berlin
Books on Demand GmbH, Norderstedt
Reha GmbH, Saarbrücken
Amazon Distribution GmbH, Leipzig
ISBN: 978-3-8381-1439-2

Imprint (only for USA, GB)
Bibliographic information published by the Deutsche Nationalbibliothek: The Deutsche Nationalbibliothek lists this publication in the Deutsche Nationalbibliografie; detailed bibliographic data are available in the Internet at http://dnb.d-nb.de.

Any brand names and product names mentioned in this book are subject to trademark, brand or patent protection and are trademarks or registered trademarks of their respective holders. The use of brand names, product names, common names, trade names, product descriptions etc. even without a particular marking in this works is in no way to be construed to mean that such names may be regarded as unrestricted in respect of trademark and brand protection legislation and could thus be used by anyone.

Publisher: Südwestdeutscher Verlag für Hochschulschriften Aktiengesellschaft & Co. KG
Dudweiler Landstr. 99, 66123 Saarbrücken, Germany
Phone +49 681 37 20 271-1, Fax +49 681 37 20 271-0
Email: info@svh-verlag.de

Printed in the U.S.A.
Printed in the U.K. by (see last page)
ISBN: 978-3-8381-1439-2

Copyright © 2010 by the author and Südwestdeutscher Verlag für Hochschulschriften Aktiengesellschaft & Co. KG and licensors
All rights reserved. Saarbrücken 2010

Identification and Characterisation of ribosomal biosynthesis pathways of two cyclic peptides from cyanobacteria

Dissertation

zur Erlangung des akademischen Grades
doctor rerum naturalium
(Dr. rer. nat.)

im Fach Biologie

eingereicht an der
Mathematisch-Naturwissenschaftlichen Fakultät I
der Humboldt-Universität zu Berlin
von

Diplom-Biologin Nadine Ziemert
geboren am 07.01.1981 in Berlin

Präsident der Humboldt-Universität zu Berlin
Prof. Dr. Dr. h.c. Christoph Markschies

Dekan der Mathematisch-Naturwissenschaftlichen Fakultät I
Prof. Dr. Lutz-Helmut Schön

Gutachter/innen: 1. Prof. Dr. Elke Dittmann
2. Prof. Dr. Wolfgang Lockau
3. Prof. Dr. Christian Hertweck

Tag der mündlichen Prüfung: 23.10.2009

Abstract

Microbial natural products represent a major source for the development of new therapeutic agents. A diverse array of compounds is produced by cyanobacteria, a heterogenous group of aerobic photoautotrophs. A variety of bioactive metabolites with potential anti-cancer, anti-microbial and anti-HIV activities have been isolated. Most of the compounds are peptides or possess peptidic structures and are usually made by large non-ribosomal assembly lines. However, a ribosomal origin has recently been demonstrated for the biosynthesis of patellamides, cytotoxic cyclic peptides produced by cyanobacterial symbionts of ascidians.

Microcystis aeruginosa NIES298 produces various peptides including microcystin, aeruginosin, microviridin and microcyclamide. For the latter two classes of peptides ribosomal biosynthesis pathways could be identified in the course of this study. The cytotoxic hexapeptide microcyclamide is formed through the activity of a set of enzymes closely related to those involved in patellamide biosynthesis. The multicyclic microviridin family of protease inhibitors are synthesised from a precursor peptide by a unique pathway involving uncharted ATP-grasp type ligases as well as an *N*-acetyltransferase and a specialised transporter peptidase. The successful expression of microviridin B in *E. coli* provides a promising base for engineering novel variants.

Screening of *Microcystis* laboratory strains and field samples revealed a wide-spread occurrence and a great natural variety for both peptide classes, raising the question of the ecological role of such small cyclic peptides. Attempting to obtain some first hints to answer that question, transcription and expression studies of biosynthetic genes were performed. Finally, this work showed that such scanning approaches could lead to the discovery of novel peptide variants and demonstrated new examples of succesful genome mining.

Zusammenfassung

Naturstoffe sind eine der wichtigsten Quellen für die Entwicklung neuer Pharmazeutika. Eine Vielzahl von bioaktiven Substanzen mit potentieller Anti-Krebs, Anti-HIV oder antimikrobieller Wirkung wurde aus der heterogenen Gruppe der photoautotrophen Cyanobakterien isoliert. Die meisten dieser Metabolite sind Peptide oder besitzen peptidähnliche Strukturen und werden nicht-ribosomal von großen, modular aufgebauten Enzymkomplexen gebildet. Vor kurzem konnte anhand der Patellamide gezeigt werden, dass zyklische Peptide auch ribosomal hergestellt werden können.

Microcystis aeruginosa NIES298 produziert eine Reihe von Sekundärmetaboliten, unter anderem die nicht-ribosomalen Peptide Microcystin und Aeruginosin. Zwei weiteren von diesem Stamm produzierten Peptiden, Microcyclamid und Microviridin B, konnten bislang noch keine Gene zugeordnet werden. In dieser Studie wurden ribosomale Biosynthesewege für beide Peptidfamilien identifiziert. Die zur Biosynthese des cytotoxischen Hexapeptids Microcyclamid notwendigen Enzyme zeigen eine hohe Ähnlichkeit zu den Patellamid-Enzymen und weisen auf ähnliche Biosynthesemechanismen hin. Ein völlig neuer Syntheseweg, in dem bis dahin unbekannte ATP-grasp-Ligasen eine Rolle spielen, konnte für den tri-zyklischen Proteaseinhibitor Microviridin gefunden werden. Die erfolgreiche heterologe Expression dieses Peptids in *E. coli* bietet die Möglichkeit ganze Bibliotheken von Microviridin-Varianten mit neuen oder verbesserten Bioaktivitäten zu konstruieren.

Die systematische Suche nach ähnlichen Biosynthesegenen in *Microcystis* Laborstämmen und Gewässerproben zeigte eine weite Verbreitung und eine große Diversität der untersuchten Peptidklassen in Cyanobakterien, und stellt die Frage nach der natürlichen Funktion dieser Metabolite. Um erste Hinweise zu erhalten, wurden Trankriptions- und Expressionsstudien der Biosynthesegene durchgeführt. Schließlich konnten, mit Hilfe des so genannten „genome-mining", neue Varianten der untersuchten Peptidklassen gefunden und aufgeklärt werden.

Table of Contents

1 **INTRODUCTION** 1
    1.1 THE IMPACT OF NATURAL COMPOUNDS 1
    1.2 SOURCES OF NATURAL COMPOUNDS 2
    1.3 NEW STRATEGIES OF DRUG DISCOVERY 5
    1.4 BIOSYNTHESIS OF SECONDARY METABOLITES: RIBOSOMAL VERSUS NONRIBOSOMAL PATHWAYS 7
    1.5 CYANOBACTERIA 10
        1.5.1 Secondary metabolites of cyanobacteria 12
        1.5.2 Patellamide biosynthesis 13
        1.5.3 Compounds of Microcystis 15
        1.5.4 Microcyclamide 16
        1.5.5 Microviridins 17
    1.6 AIMS OF THIS STUDY 19

2 **MATERIALS AND METHODS** 20
    2.1 MATERIALS 20
        2.1.1 Bacterial strains 20
        2.1.2 Chemicals 21
        2.1.3 Kits 24
        2.1.4 Radiochemicals 24
        2.1.5 Enzymes 25
        2.1.6 Filters and Membranes 25
        2.1.7 Marker 25
        2.1.8 Antibodies 26
        2.1.9 Nucleic Acids 26
    2.2 METHODS 28
        2.2.1 Cultivation of bacteria 28
        2.2.2 Molecular biological techniques 28
        2.2.3 Protein biochemical methods 33
        2.2.4 Immunofluorescence Microscopy 39
        2.2.5 Phylogenetic analysis 40

3 RESULTS ... 42

3.1 MICROCYCLAMIDES ... 42
- 3.1.1 Identification of microcyclamide biosynthesis genes in Microcystis aeruginosa NIES298 ... 42
- 3.1.2 Comparative analysis of the mca genes ... 44
- 3.1.3 Heterologous expression of microcyclamide ... 48
- 3.1.4 Transcription of the mca genes in M. aeruginosa NIES298 ... 48
- 3.1.5 An orphan microcyclamide-like gene cluster in M. aeruginosa PCC7806 ... 49
- 3.1.6 Variability of microcyclamides in Microcystis ... 53

3.2 MICROVIRIDINS ... 56
- 3.2.1 In search of the biosynthesis pathway ... 56
- 3.2.2 Microviridin biosynthesis gene clusters in Microcystis ... 58
- 3.2.3 Heterologous expression of microviridins ... 60
- 3.2.4 Characterisation of the microviridin ligases ... 62
- 3.2.5 Variety of microviridins ... 67
- 3.2.6 Preliminary characterisations of the putative ABC transporter MdnE ... 71
- 3.2.7 Microviridin expression studies in cyanobacteria ... 73

4 DISCUSSION ... 76

4.1 MICROCYCLAMIDE BIOSYNTHESIS IN *MICROCYSTIS* ... 77
- 4.1.1 A patellamide-like biosynthesis of microcyclamides ... 77
- 4.1.2 Shedding light on the precursor ... 80
- 4.1.3 Diversity of microcyclamides ... 81
- 4.1.4 Towards the role of microcyclamides ... 84

4.2 MICROVIRIDIN BIOSYNTHESIS – A NOVEL RIBOSOMAL PATHWAY ... 87
- 4.2.1 The microviridin ligases ... 88
- 4.2.2 The N-acetyltransferase ... 91
- 4.2.3 Which role does the ABC transporter play? ... 92
- 4.2.4 Microviridins – another diverse family of cyanobacterial peptides ... 94
- 4.2.5 Possible applications in bioengineering ... 96
- 4.2.6 Functional hypotheses for microviridins ... 98

4.3 GENERAL CONSIDERATION ABOUT POSSIBLE FUNCTIONS OF CYANOBACTERIAL SECONDARY METABOLITES 100

5 REFERENCES 102

APPENDIX 114

ABBREVIATIONS 115

EIGENSTÄNDIGKEITSERKLÄRUNG 118

ACKNOWLEDGEMENT 119

1 Introduction

1.1 The impact of natural compounds

Natural products have always been and, to this day, still are omnipresent in the everyday life of all humankind. Although at times not readily identified as natural products by the average person, today, everybody can appreciate the benefits brought about by substances such as caffeine or penicillin (fig. 1). Natural products are defined as compounds, which have biological activities and are derived from natural sources, such as plants, animals and microorganisms (Baker et al., 2007). Most natural product compounds are secondary metabolites from plants and microbes. Ancient cultures already knew the effects of certain plants in fighting infections and diseases and long before the discovery of microorganisms, they were used to produce alcohol, vinegar and cheese (Demain, Fang, 2000). With the discovery of penicillin by Alexander Fleming in 1929 (Fleming, 1929) and its large scale production during World War II, the golden era of natural compounds began. The discovery of more antibiotics such as streptomycin, gentamycin and tetracycline revealed the potential of microbial secondary metabolites in fighting bacterial and fungal infections. In the 1960s first efforts began to broaden the scope of natural products and since then many compounds with antiparasitic, anticancer, anti-HIV, enzyme inhibitory or immunosuppressive properties were discovered (Zerikly, Challis, 2009). By 1990, about 80% of drugs were either natural products or analogs inspired by them (Li, Vederas, 2009).

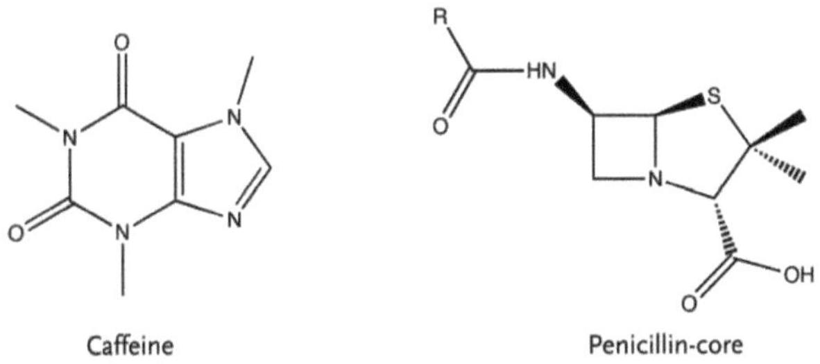

Caffeine   Penicillin-core

Fig.1 Structure of caffeine and the penicillin core unit as examples for known natural compounds.

With the development of new techniques in genetic engineering and combinatorial chemistry, natural molecules have less frequently served as final applications but emerged as lead molecules for directed manipulations. This altered strategy promises significant enhancements in product performance and was made possible to a large extent by the rapidly growing understanding of the basic mechanisms underlying the biosynthesis of natural compounds. In the case of penicillin biosynthesis, both biosynthetic genes and pathways as well as complex regulatory networks were identified (Brakhage, 1997; Macdonald, Holt, 1976). Although we are still far from a full mechanistic understanding of natural product biosynthesis, findings like these have not only proven to be invaluable for the industrial production of bioactive substances but have greatly advanced our knowledge of the ecological roles of these compounds in the environment. Whereas it was once popular to think that secondary metabolites were laboratory artefacts, it is known today that these compounds serve not only as competitive weapons against bacteria, fungi, amoebae, plants, insects and large animals, but also as metal transporting agents, sexual hormones, differentiation effectors and communication signals in quorum sensing systems or symbiosis between microbes and plants (Demain, Fang, 2000). On the basis of the widespread occurrence alone it can be presumed that secondary metabolites fulfil essential functions in nature for the organisms that produce them.

1.2 Sources of natural compounds

Many natural products are made by plants. Diazetylmorphin (1898)(fig.2) and acetylsalicic acid (1899) were among the first compounds to be commercialised as pharmaceuticals, and today plants remain a major source for drug discovery, with 91 metabolites in clinical trials as of late 2007 (Li, Vederas, 2009; Potterat, Hamburger, 2008). There are also examples for natural compounds derived from animals. Magainins are antimicrobial peptides, isolated from the skin of the frog *Xenopus* (Zasloff, 1987). A peptide originally discovered in a tropical cone snail: ziconotide (Prialt; Elan Pharmaceuticals), was approved for the treatment of pain.

Another drug — the antitumour compound trabectedin (Yondelis; PharmaMar) from a tropical sea-squirt (fig.2) — is used for the treatment of soft-tissue sarcoma (Molinski et al., 2009). However, microbial natural products represent a major source for new drug candidates. Bioactive compounds from fungi contain not only antibiotics, such as the already mentioned penicillin or cephalosporin, but also potent toxins such as amatoxins (e.g. α-amanitin in fig. 2) and phalloidins (Vetter, 1998). The cytochalasans comprise a diverse group of mycotoxins with e.g. antibiotic, antiviral or antitumor properties (Binder, Tamm, 1973; Turner, Carter, 1972). Filamentous soil bacteria are traditionally main sources for natural compounds. Actinomycetes produce 74% of all described antibiotics (Demain, Fang, 2000). From myxobacteria more than 100 different basic compounds and approximately 500 structural variants have been characterised over the last two decades (Wenzel, Müller, 2007). Members of the bacterial genera *Bacillus* and *Streptococcus* are also quite active in this respect.

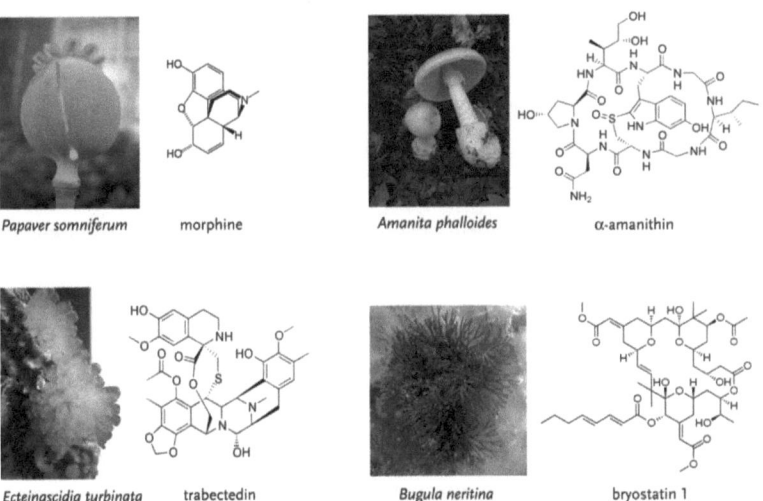

Fig. 2 Examples of organisms producing natural products.
(Images from http://en.wikipedia.org/wiki/)

In the late 1960s, the search for novel metabolites extended from terrestrial organisms to aquatic environments (Jensen, Fenical, 1994). The world's oceans and seas, which cover over 70% of the earth's surface, have shown to be a prolific source for bioactive metabolites. Since the 1970s more than 15,000 structurally diverse compounds with different bioactivities have been discovered (Baker et al., 2007). These include two anticancer agents, such as bryostatin 1 (fig.2) (Hennings et al., 1987) and didemnin B (Rinehart et al., 1981), which are currently in clinical trials.

In the last years symbiotic interactions between organisms have become of great interest to natural product researchers. With regard to secondary metabolism, these life forms are enormously productive (König et al., 2006). One famous example is Paclitaxel, also known as taxol, a mitotic inhibitor used in cancer chemotherapy. It has been harvested from the dried and inner bark of *Taxus brevifolia* also called Pacific yew (Walsh, Goodman, 1999) (fig. 3A). In 1993 taxol was discovered to be produced in a newly described endophytic fungus (Stierle et al., 1993). Another anti-tumor compound, Rhizoxin, was originally isolated from a pathogenic plant fungus (*Rhizopus microsporus*). It has been shown that this phytotoxin is not produced by the fungus, but by symbiotic bacteria of the genus *Burkholderia* that reside within the fungal cytosol (fig. 3B) (Partida-Martinez, Hertweck, 2007). A frequently encountered hypothesis is that most marine compounds are secondary metabolites of invertebrate-associated bacteria and fungi (König et al., 2006). Especially sponges are hosts of highly complex symbiont communities and excellent natural product sources (Piel, 2009).

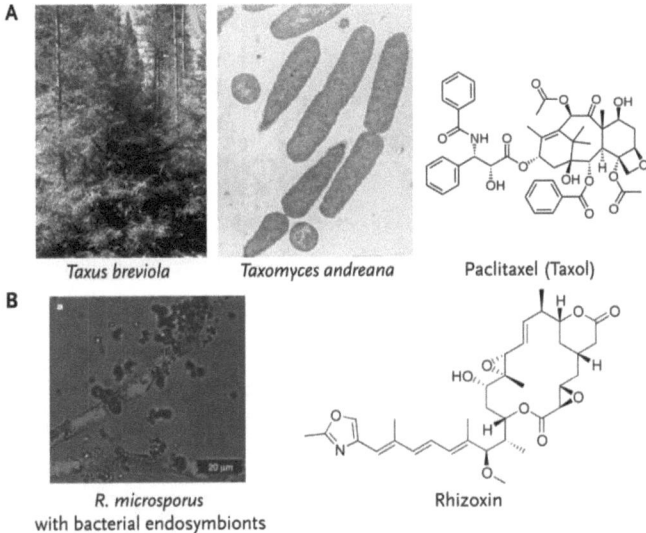

Fig. 3 Symbiotic relationships and their natural products.
A *Taxus breviola* (http://www.forestryimages.org) and its endophytic fungus *Taxomyces andreana* (Weeks, Alcamo, 2007), which produces Paclitaxel.
B Bacterial endosymbionts in the cytoplasm of *R. microsporus* ATCC 62417 (Partida-Martinez, Hertweck, 2005).

1.3 New strategies of drug discovery

Traditionally, the discovery of natural products was based on screening of various biological sources for different bioactivities. Since most symbionts and free-living bacteria are as yet unculturable, advances in culture-independent techniques are important to elucidate biosynthetic origins of natural compounds. The large quantity of publicly accessible DNA sequences, metagenomic libraries and heterologous expression methods makes it possible to discover new promising natural products. The ability to rationally alter natural-product structures through the genetic modification of their biosynthetic machinery, also known as combinatorial biosynthesis, has led to the production of libraries of "non-natural" natural products and has broadened the variety of compounds (Wilkinson, Micklefield, 2007).

The increasing number of fully sequenced genomes and the improved understanding of the genetics and enzymology facilitates the identification and analysis of cryptic biosynthetic gene clusters (Jenke-Kodama, Dittmann, 2009a; Wilkinson, Micklefield, 2007; Zerikly, Challis, 2009). Genomic mining reveals that even well known sources can comprise cryptic gene clusters, which no products could be assigned to. One of the first sequenced microbial genomes was that from *Streptomyces coelicolor* in 2001. Although studied extensively and often regarded as a model streptomycete, more gene clusters encoding for natural products have been found than there were known natural products of the organism (Bentley *et al.*, 2002). Different approaches aim to identify products to those cryptic gene clusters. Bioinformatic analysis can predict physiochemical properties of the product, facilitating the identification of the respective compound (Jenke-Kodama, Dittmann, 2009a). If substrates of the encoded biosynthetic enzymes are known, a directed feeding of isotope-labelled precursors can guide the detection of products, or *in vitro* analysis of the purified enzymes and their catalysed reactions can be done (Zerikly, Challis, 2009).

Nature provides much more variety in bioactive compounds, than we have discovered by now. New techniques and more insights into their biosynthesis pathways will help us to meet the growing requirements in developing new drugs and combat common diseases such as cancer and the increasing bacterial resistance to currently used antibiotics.

1.4 Biosynthesis of secondary metabolites: ribosomal versus nonribosomal pathways

During the last two decades, enormous progress has been made in elucidating the biosynthesis of hundreds of secondary metabolites, mostly from microorganisms (Donadio et al., 2007). The majority of these metabolites are linear or cyclic peptides comprised of many unusual as well as proteinogenic amino acids containing a number of modifications. A major part of these oligopeptides were shown to be products of non-ribosomal assembly lines (Finking, Marahiel, 2004). Non-ribosomal peptide synthetases (NRPS) could be identified for the biosynthesis of common antibiotic families such as the penicillins and cephalosporins (Schofield et al., 1997). NRPS are large multi-enzyme complexes (fig. 4). They have a modular structure, with each module being responsible for the activation, thiolation, modification and condensation of one specific amino acid (Marahiel et al., 1997). Each module consists of a number of domains two of which are commonly referred to as A (adenylation) and C (condensation). The A domain activates a specific amino acid (analogous to a t-RNA) and transfers it to the PCP (peptidyl carrier protein) which holds on to the growing peptidyl as a thioester. The C domain forms a peptide bond between the next amino acyl and the peptidyl unit.

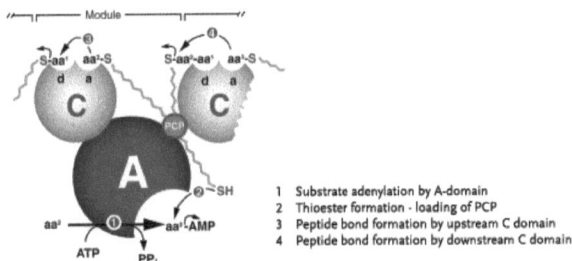

Fig. 4 **Overview of the composition and workflow of NRPS** (Weber, Marahiel, 2001). Schematic representation of peptide biosynthesis in an NRPS minimal module.

Modifying domains for epimerisation, heterocyclisation or oxidation could be additionally integrated. Polyketide side chains can be incorporated by modular polyketide synthases, which are functionally related to NRPS. Contrary to peptides and their amino acid building blocks, polyketides are assembled from acyl units (Hertweck, 2009).

Characterisation of ribosomal small modified peptides such as the conopeptides from molluscs (Olivera, 2006), cyclotides from plants (Trabi, Craik, 2002) and the fungal toxins amanitins and phallacidins (Hallen et al., 2007) revealed that there are biosynthetic pathways independent from nonribosomal peptide synthetase systems. Many of the modifications commonly thought of being confined to nonribosomal assembly pathways, could be also shown in ribosomal peptides (RPs) (McIntosh et al., 2009).

All known RPs are derived from relatively short precursor proteins that have been translated and then modified. That modification includes at least one proteolytic step, where cleavage of the active peptide from so called leader or signal peptides occurs. One well-studied group of RPs in bacteria are the bacteriocins, including the microcins of *Escherichia coli* (fig. 5) and the lanthionine-containing lantibiotics of gram-positive bacteria. Members of this group exhibit antibiotic activities against other bacteria and were shown to increase the permeability of cell membranes and to inhibit DNA gyrases and RNA polymerases (Breukink, 2006; Jack, Jung, 2000). Lantibiotics such as nisin (fig. 5) are also used for food preservation (Delves-Broughton et al., 1996). Besides their role in defence against other bacteria, lantibiotics of gram-positive bacteria were shown to play a role in the cell-cell signalling of bacteria (Kleerebezem et al., 1997). Most bacteriocins and lantibiotics contain a characteristic *N*-terminal leader sequence with a double glycine motif, which is cleaved by a dedicated ABC transporter concomitant with translocation across the membrane (Michiels et al., 2001).

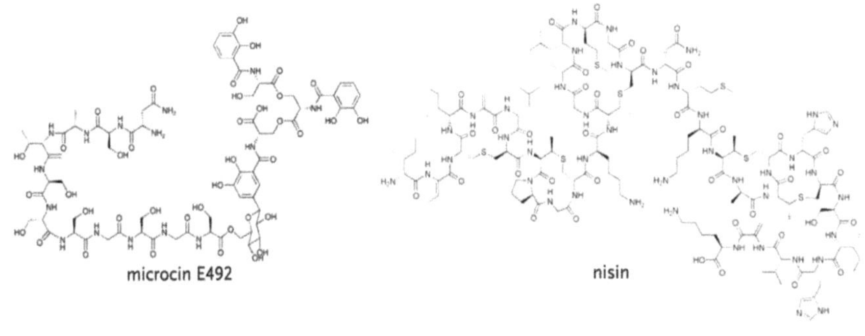

Fig. 5 Structures of selected bacteriocins. (McIntosh et al., 2009)

According to their modular structure, NRPS and modular PKS or NRPS/PKS hybrid systems have been proven to be very amenable to engeneering (Menzella, Reeves, 2007). Over the last 15 years various genetic modifications, including replacement of domains, insertion, deletion or replacement of entire modules have been applied to obtain new variants of natural compounds. However, many engeneering attempts have given inactive or very inefficient enzymes. Although non-ribosomal assembly lines are unique in their ability to incorporate non-proteinogenic amino acids, fatty acids or polyketides, genetic engineering and heterologous expression of these giant enzymes is hindered by their size and their need of certain enzymes such as 4´ phospopantetheinyl transferases (PPTases) and type II thioesterases, which are uncommon in hosts like *E. coli* (Doekel et al., 2002). Furthermore, biocombinatorial manipulation of NRPSs is impeded by communication-mediating domains (COM), which tolerate only specific arrangement of different modules (Hahn, Stachelhaus, 2004). It requires about 3000 nucleotides to add one amino acid in the NRP system, while only three are required for an RP codon. The characterisation of the various tailoring enzymes of ribosomal biosynthesis pathways can provide useful tools for biotechnological applications. Although less celebrated as potential medicines, ribosomal peptides can expand the chemical diversity of useful natural products and provide tools for their biotechnological use, as the example of patellamides demonstrates (chapter 1.5.2).

1.5 Cyanobacteria

With fossil records dating back to nearly 3.5 billion years ago, cyanobacteria (also blue-green algae) belong to the oldest extant organisms on earth (Schopf, 1993). The majority of them are aerobic photoautotrophs, which possess chlorophyll *a* and perform oxygenic photosynthesis associated with photosystems I and II, an activity that resulted in the enrichment of the planetary atmosphere in oxygen, so that about 1.5 billion years ago our vital atmosphere had been created (Des Marais, 1991). Cyanobacteria get their name from the phycobilin pigment phycocyanin, which leads to the bluish colour of the organism. According to the endosymbiont theory, ancient cyanobacteria are the ancestors of plastids and therefore an important part of the evolution of eukaryotic phototrophic cells (Giovannoni *et al.*, 1988).

Today cyanobacteria form a huge and heterogeneous group of prokaryotes. They comprise unicellular, colonial and multicellular filamentous forms (fig. 6) (Stanier, Cohen-Bazire, 1977). Certain species are able to develop differentiated cell forms such as heterocysts for nitrogen fixation or akinetes as specialised resting cells. Many aquatic cyanobacteria are able to form gas vesicles, which make the cells buyont and enable their migration through the water column. The prominent habitats of cyanobacteria are limnic and marine environments (Mur *et al.*, 1999). They flourish in water that is salty, brackish or fresh, in cold and hot springs, and arctic and antarctic lakes. They are able to colonise infertile substrates such as volcanic ash, desert sand and rocks and form symbiotic associations with animals and plants (Whitton, Potts, 2000). Among their symbiotic partners are species from fungi, bryophytes, pteridophytes, gymnosperms and angiosperms (Rai, 1990).

Cyanobacteria provide an extraordinarily wide-ranging contribution to human affairs in everyday life. They are important primary producers. Marine phytoplankton accounts for nearly 50 % of the net primary productivity of the biosphere (Field *et al.*, 1998), the nitrogen-fixing species contribute globally to soil and water fertility (Montoya *et al.*, 2004) and the use of cyanobacteria in food (*Spirulina*) and fuel production holds promising potential for the future (Angermayr *et al.*, 2009).

However, cyanobacteria also cause considerable concerns for human health. Under certain conditions, especially where waters are rich in nutrients and exposed to sunlight, cyanobacteria may multiply to high densities – so called blooms. Various types of cyanobacteria such as *Microcystis, Anabaena* and *Planktothrix* frequently form toxic blooms in freshwater lakes (Mur *et al.*, 1999). Nutrient overenrichment of waters and rising temperatures now exacerbated by global warming promote the growth of cyanobacteria as harmful algal blooms and lead to a dramatic decrease of biodiversity in these ecosystems (Paerl, Huisman, 2008).

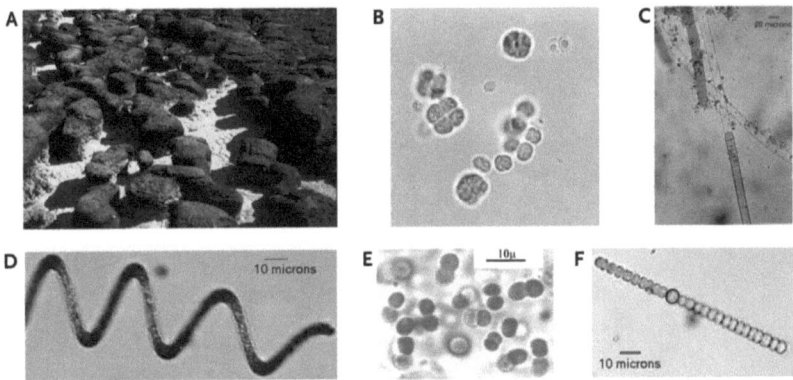

Fig. 6 The variety of cyanobacteria.
A Stromatolites, layered fossil structures formed in shallow water by the trapping, binding and cementation of sedimentary grains by ancient cyanobacteria. B Colonies of **Cyanosarcina** sp. C Filaments of **Lyngbya** sp. D *Arthrospira* sp. E *Synechocystis* sp. F *Anabaena* sp. (Images from http://www-cyanosite.bio.purdue.edu/images/images.html)

1.5.1 Secondary metabolites of cyanobacteria

A diverse array of metabolites are found in cyanobacteria. Two prominent molecules produced by marine cyanobacteria, curacin A and dolastatin, have been in preclinical and clinical trials as potential anticancer drugs (Gerwick et al., 2001). Freshwater cyanobacteria are well known for the production of lethal toxins. Due to their adverse effects on higher organisms, compounds such as the hepatotoxins microcystin and cylindrospermopsin or the neuroxins anatoxin-a and saxitoxin (fig. 7) have given cause for serious concern of water authorities worldwide (Chorus et al., 2000).

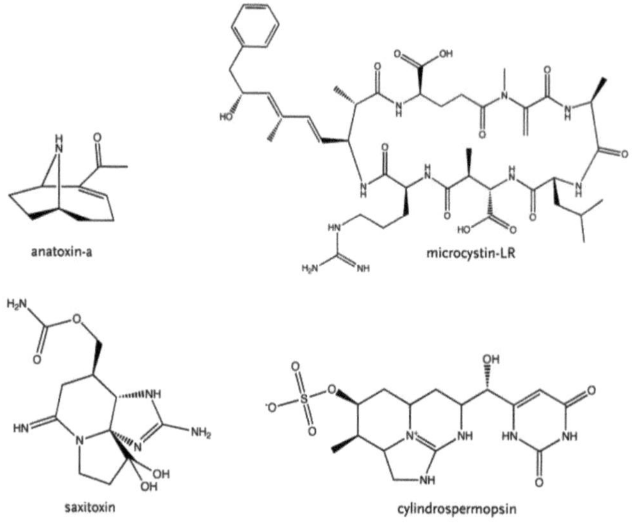

Fig. 7 Known toxins from cyanobacteria.

However, cyanobacteria are not only known to produce toxins, but a multitude of compounds covering a broad spectrum of bioactivities (Tan, 2007; Welker, von Döhren, 2006). Structures with anticancer, antibacterial, antifungal and protease inhibitory effects have been found (Namikoshi, Rinehart, 1996).

Anti-HIV activities were shown for cyanovirin-N and microvirin, two lectins isolated from *Nostoc ellipsosporium* (Mori et al., 1998) and *Microcystis aeruginosa* (Kehr et al., 2006), respectively. Anti-malaria screenings have led to the isolation of gallinamides (Linington et al., 2009) and venturamides (Linington et al., 2009) from marine cyanobacteria.

A majority of these metabolites, in particular those that were isolated from planktonic freshwater cyanobacteria belonging to the genera *Microcystis*, *Planktothrix*, *Nostoc*, and *Anabaena*, can be classified as peptides or possess peptidic substructures often comprising highly modified amino acid moieties. So far, more than 600 peptides are described from various taxa (Welker, von Döhren, 2006). The majority of these peptides, such as the hepatotoxin microcystin or the protease inhibitors aeruginosin and anabaenopeptolide, were shown to be produced by nonribosomal peptide synthetase assembly lines (Ishida et al., 2007; Rouhiainen et al., 2000; Tillett et al., 2000). At the beginning of this work, only one example for a ribosomal peptide produced by cyanobacteria was known – the patellamides (Schmidt et al., 2005).

1.5.2 Patellamide biosynthesis

The patellamide family of peptides are cyclic pseudosymmetrical octapeptides (fig. 8C), which are characterised by the presence of thiazole and oxazole moieties. Although nonribosomal biosynthesis was anticipated for the formation of these peptides, heterologous expression of a microcin-like gene cluster (fig. 8C) discovered in the genome of the symbiotic cyanobacterium *Prochloron didemni* (fig. 8B) unambiguously showed that these peptides are produced by a ribosomal pathway (Long et al., 2005; Schmidt et al., 2005).

In a more recent study, the patellamide biosynthetic pathway could be used as a template for the design of a highly flexible expression platform for the production of libraries of cyclic peptides (Donia et al., 2006).

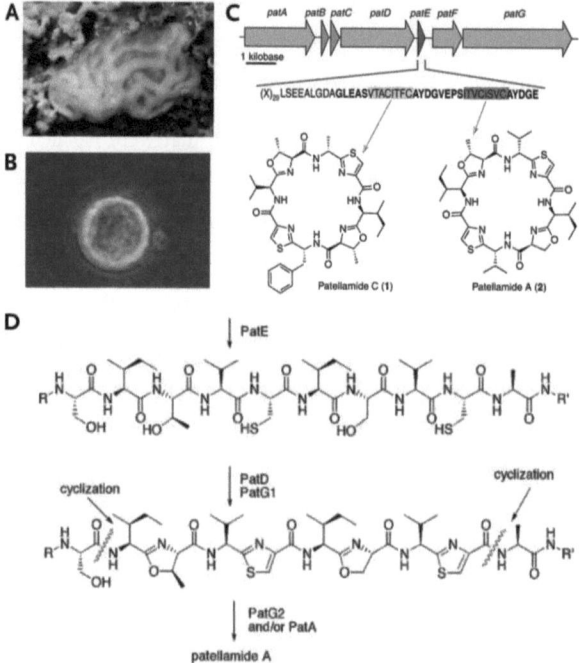

Fig. 8 Biosynthesis of patellamides in Prochloron, a cyanobacterial symbiont of an ascidian.
A The ascidian *Lissoclinum patella*. B Single cell of *Prochloron didemni*. (Schmidt et al., 2005)
C The pat cluster encodes seven coding sequences: patA– patG. The patE gene (red), encodes the peptide sequence for patellamide A (2, sequence in green) and patellamide C (1, sequence in yellow). Essential genes for in vivo production of patellamides are shown in orange. Putative start and stop recognition sequences are shown in bold. (Donia et al., 2006)
D Proposed pathway to patellamides showing the route to patellamide A. (Schmidt et al., 2005)

1.5.3 Compounds of *Microcystis*

The freshwater cyanobacterium *Microcystis* (fig. 9) is one of the most common bloom-forming cyanobacterial species with a cosmopolitan distribution. *Microcystis* is characterised as having a coccoid cell shape, gas vesicles, a tendency to form aggregates or colonies and an amorphous mucilage or a sheath (Otsuka *et al.*, 2000). It produces a variety of different toxic and nontoxic metabolites, such as microcystins, microginins, cyanopeptolins, aeruginosins, microviridins and microcyclamide (Czarnecki *et al.*, 2006; Ishida *et al.*, 2007; Ishida *et al.*, 2000; Ishitsuka *et al.*, 1990; Okino *et al.*, 1993; Tillett *et al.*, 2000). Whereas non-ribosomal assembly lines to most of these peptides could be assigned for, the biosynthesis genes of microcyclamide and the microviridins were still elusive at the beginning of this study.

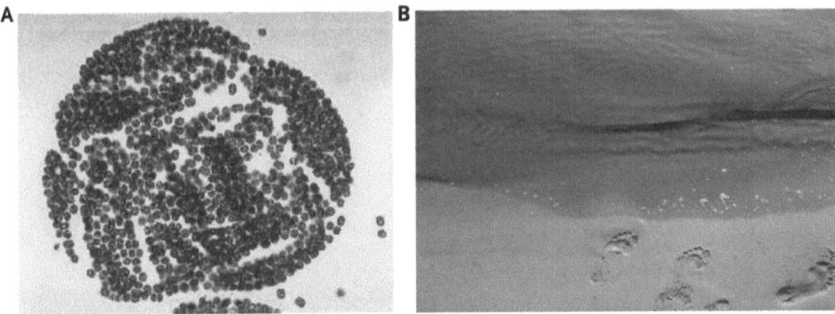

Fig. 9 The cyanobacterium *Microcystis*.
A Colony of *Microcystis* spec. (http://www-cyanosite.bio.purdue.edu/images/images.html)
B *Microcystis* bloom at Lake Wannsee in Berlin, Germany.

1.5.4 Microcyclamide

Microcyclamide is a cytotoxic cyclic hexapeptide produced by the cyanobacterium *Microcystis aeruginosa* NIES298 (Ishida *et al.*, 2000) (fig. 10). It contains three five-membered heterocycles (two thiazoles and one methyloxazole) and therefore shows some structural similarity to the aforementioned patellamides from *Prochloron didemni*. Various cyclic peptides with thiazole and oxazole moieties have been described in free-living and symbiotic cyanobacteria, including nostocyclamide (Jüttner *et al.*, 2001) (fig. 10), tenuecyclamide (Baker *et al.*, 2007) (fig. 10), venturamides (Linington *et al.*, 2007) and dendroamides (Ogino *et al.*, 1996). Although the naming of this peptide class is very incoherent, they have been summarised in 2006 as cyclamides (Welker, von Döhren, 2006). In all cyclamides thiazole/oxazole units occur in alternation with unmodified amino acids to form a cyclic hexapeptide. The variety of structures is reflected in an equally large variety of bioactivities, such as antibacterial, cytotoxic and antimalarial activities. Although a patellamide - like biosynthesis pathway (chapter 1.3) has been assumed for these peptides, no respective genes have been found, before this study has begun. Furthermore, their biological function and evolution are still under debate.

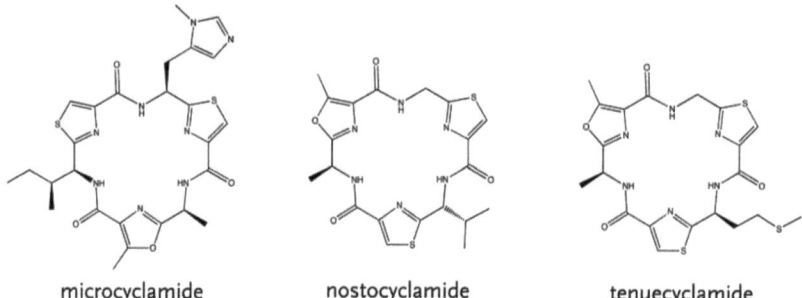

microcyclamide   nostocyclamide   tenuecyclamide

Fig. 10 Structures of known cyclamides.

1.5.5 Microviridins

Some of the most remarkable cyanobacterial toxins belong to the microviridin family of depsipeptides (peptides bearing ester bonds) (Ishitsuka et al., 1990; Murakami et al., 1997; Okino et al., 1995; Rohrlack et al., 2003). These largest known cyanobacterial oligopeptides are produced by a number of freshwater species such as *Microcystis*, *Planktothrix* and *Nostoc*. Microviridins are characterised by their multicyclic architecture that results from intramolecular ω-ester and ω-amide bonds (fig. 11). The main peptide ring consists of seven amino acids with an ester bond between the 4-carboxy group of aspartate and the hydroxy group of threonine and a peptide bond between the 6-amino group of lysine and the 4-carboxy group of glutamate (Welker, von Döhren, 2006). The characteristic core motif for microviridins is the five amino acid sequence KYPSD, variations are primarily due to substitutions in the side chain. The *N*-terminal amino acid is commonly acetylated. Microviridins are potent and very specific inhibitors of various types of proteases. While microviridin B has high therapeutic potential as an elastase inhibitor in the treatment of lung emphysema (Murakami et al., 1997; Okino et al., 1995), another representative of this peptide family, microviridin J, was shown to be toxic for *Daphnia* spp., a key group of organisms in freshwater ecosystems (Rohrlack et al., 2004). Table 1 shows selected representatives of the microviridin peptides and their inhibitory activities.

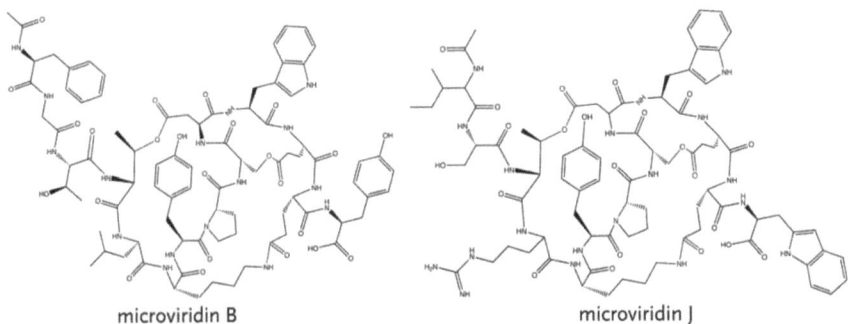

Fig. 11 Structures of microviridin B and J.
Crosslinks from ω-amide and ω-ester bonds are highlighted in red.

Tab. 1: Selected Microviridins and their activities. Conserved amino acids are in red.

Microviridin	Amino acid sequence	genus	Inhibitory activity	Ref.
A	YGGTFKYPSDWEEY	*Microcystis*	Tyrosinase	(Ishitsuka et al., 1990)
B	FGTTLKYPSDWEEY	*Microcystis*	Elastase	(Okino et al., 1995)
C	FGTTLKYPSDWEEY	*Microcystis*	Elastase	(Okino et al., 1995)
D	YGNTMKYPSDWEEY	*Planktothrix*	Elastase/chymotrypsin	(Shin et al., 1996)
E	FSTYKYPSDFEDF	*Planktothrix*	Elastase/chymotrypsin	(Shin et al., 1996)
F	FSTYKYPSDFEDF	*Planktothrix*	Elastase	(Shin et al., 1996)
G	YPQTLKYPSDWEEY	*Nostoc*	Elastase	(Murakami et al., 1997)
H	YPQTLKYPSDWEEY	*Nostoc*	Elastase	(Murakami et al., 1997)
J	ISTRKYPSDWEEW	*Microcystis*	trypsin/chymotrypsin	(Rohrlack et al., 2003)

Depsipeptides are generally synthesised by non-ribosomal peptide synthetase (NRPS) assembly lines, as in the cyanobacterial cyanopeptolin biosynthetic pathway. However, despite extensive studies, no candidate genes for microviridin biosynthesis were identified by now (Nishizawa et al., 2007) (Dittmann, Kaebernick, personal communications). Because microviridins are solely composed of proteinogenic L-α-amino acids, we hypothesised that they derive from a ribosomal biosynthesis.

1.6 Aims of this study

Cyanobacteria are prolific producers of secondary metabolites with various bioactivities. Identification and characterisation of the biosynthetic pathways provide useful tools for their biotechnological approaches. The freshwater cyanobacterium *Microcystis aeruginosa* is known to produce a variety of bioactive compounds, such as microcystin, cyanopeptolin and aeruginosin (Welker, von Döhren, 2006). For the majority of these compounds corresponding biosynthesis genes could be identified (Czarnecki *et al.*, 2006; Ishida *et al.*, 2007; Tillett *et al.*, 2000). Usually large NRPS or NRPS/PKS hybrid enzyme complexes are responsible for their biosynthesis. *M. aeruginosa* NIES298 produces, amongst others, two compounds, which no biosynthesis genes could be assigned to: The cytotoxic compound microcyclamide, a cyclic hexapeptide containing thiazole and oxazole moieties, and the tricyclic depsipeptide microviridin B, a potent elastase inhibitor. This study aims at elucidating the biosynthesis pathways of both cyclic peptides and obtaining first insight into the molecular mechanisms underlying their synthesis. Analysis of genomic data and screening of field samples are thought to clarify the abundance and variety of these peptide families and to help guiding the discovery of novel variants. Expression and transcription studies may provide first ecological insights and could be valuable for elucidating the possible functional role of these metabolites.

2 Materials and Methods

2.1 Materials

2.1.1 Bacterial strains

2.1.1.1 Microcystis

Microcystis aeruginosa strains PCC7806 and NIES298 were obtained from the Pasteur Culture Collection of Cyanobacteria (Institute Pasteur, Paris, France) and the National Institute of Environmental Studies (Tsukuba, Japan), respectively. *Microcystis aeruginosa* UWOCC MRC originates from the University of Wisconsin Culture Collection (Oshkosh, USA). To investigate the variability of microcyclamide and microviridin a set of DNA from different *Microcystis* strains was analysed (tab. 2).

Tab. 2: *Microcystis* strains used in this study and their origin

Microcystis strains	Geographic origin	References
HUB5.3	Lake Pehlitzsee, Germany	(Leikoski et al., 2009)
Izancya 5	Lake Mira, Portugal	(Leikoski et al., 2009)
199	Lake Rusutjärvi, Finland	(Leikoski et al., 2009)
269	River Raisionjoki, Finland	(Leikoski et al., 2009)
NIES100	Lake Kasumigaura, Japan	(Leikoski et al., 2009)
NIES101	Lake Suwa, Japan	(Cadel-Six et al., 2008)
NIES102	Lake Kasumigaura, Ibaraki, Japan	(Cadel-Six et al., 2008)
NIES298	Lake Kasumigaura, Japan	(Cadel-Six et al., 2008)
Nies843	Lake Kasumigaura Ibaraki, Japan	(Kaneko et al., 2007)
PCC7806	Braakman Reservoir, The Netherlands	(Cadel-Six et al., 2008)
PCC9354	Little Rideau lake, Ontario, Canada	(Cadel-Six et al., 2008)
PCC9603	Okesund dam, Sweden	(Cadel-Six et al., 2008)
PCC9804	Canberra, Australia	(Cadel-Six et al., 2008)
PCC9805	Canberra, Australia	(Cadel-Six et al., 2008)
PCC9812	Lake Mendota, Wisconsin	(Cadel-Six et al., 2008)
UWOCC MRC	Malpas Dam, Armidale, Australia	(Kaebernick et al., 2001)

2.1.1.2 Other cyanobacteria

Anabaena PCC7120 also known as *Nostoc* PCC7120 was obtained from the Pasteur Culture Collection of Cyanobacteria (Institute Pasteur, Paris, France).

2.1.1.3 Escherichia coli

The *Escherichia coli* strains XL-1 Blue (Stratagene) and TOP10 (Invitrogen, Karlsruhe) were used to amplify recombinant plasmid vectors. Heterologous expression was performed in the strain *E. coli* BL21 (Novagen). *E. coli* strain EPI300-T1R was used for fosmid library construction.

2.1.2 Chemicals

2-mercaptoethanol	C. Roth, Karlsruhe
Acetic acid	C. Roth, Karlsruhe
Acetone	C. Roth, Karlsruhe
Acetonitril "HPLC Gradient Grade"	C. Roth, Karlsruhe
Acrylamide/Bisacrylamide (37.5:1)	C. Roth, Karlsruhe
Agar, washed	Difco, Detroit
Agarose	Biozym Diagnostik, Hameln
Ampicillin	Roche Diagnostics, Mannheim
APS	C. Roth, Karlsruhe
ATP	MBI Fermentas, St. Leon-Rot
Bacto-Agar	Difco, Detroit

Bacto-Trypton	Difco, Detroit
Boric acid	C. Roth, Karlsruhe
Bovine serum albumin, fraction V	C. Roth, Karlsruhe
Bromophenole blue	SERVA Feinbiochemika, Heidelberg
Chloramphenicol	Roche Diagnostics, Mannheim
Chloroform/Isoamylalcohol 24:1	C. Roth, Karlsruhe
Coomassie staining "Roti-Blue"	C. Roth, Karlsruhe
Cyano-agar	Difco, Detroit
Dithiothreitol	C. Roth, Karlsruhe
dNTP	MBI Fermentas, St. Leon-Rot
DTNB	C. Roth, Karlsruhe
EDTA	C. Roth, Karlsruhe
Ethanol	C. Roth, Karlsruhe
Ethidium bromide	C. Roth, Karlsruhe
Formaldehyde 37%	C. Roth, Karlsruhe
GelCode Blue Stain Reagent	Pierce, Rockford
Glycerol	C. Roth, Karlsruhe
Glycine	C. Roth, Karlsruhe
HEPES	Amersham Pharmacia, Freiburg
Hydrochloric acid	C. Roth, Karlsruhe
IPTG	C. Roth, Karlsruhe
Isopropanol	C. Roth, Karlsruhe

Magnesium chloride	C. Roth, Karlsruhe
Methanol	C. Roth, Karlsruhe
N-Propylgallate	C. Roth, Karlsruhe
Phenole/Chloroform/Isoamylalcohol 25:24:1	C. Roth, Karlsruhe
PMSF	SERVA Feinbiochemika, Heidelberg
Potassium chloride	C. Roth, Karlsruhe
Skim milk powder, "Fluka"	Sigma-Aldrich Chemie, Buchs
Sodium acetate	C. Roth, Karlsruhe
Sodium chloride	C. Roth, Karlsruhe
Sodium dihydrogenphosphate	C. Roth, Karlsruhe
Sodium dodecylsulfate (SDS)	SERVA Feinbiochemika, Heidelberg
Sodium hydrogenphosphate	C. Roth, Karlsruhe
Sodium hydroxide	C. Roth, Karlsruhe
TEMED	C. Roth, Karlsruhe
trifluoroacetic acid	Sigma-Aldrich Chemie, Buchs
Tris	C. Roth, Karlsruhe
Tween 20	Sigma-Aldrich Chemie, Buchs
Urea	ICN Biochemicals, Irvine
X-Gal	C. Roth, Karlsruhe
Yeast extract	Difco, Detroit

2.1.3 Kits

Bio-Rad Protein Assay	BioRad, München
CopyControl™ Fosmid Library Production Kit	Epicentre Technologies, Madison
Hexalabel DNA Labeling	MBI Fermentas, St. Leon-Rot
Jetsorb „Gel Extraction Kit"	Genomed, Löhne
Nickel-NTA-Superflow	Qiagen, Hilden
PCR Cloning Kit pDrive	Qiagen, Hilden
PCR Purification Kit	Qiagen, Hilden
Plasmid Mini Prep	Qiagen, Hilden
SuperSignal West Pico	Pierce, Rockford
Taq DNA-Polymerase PCR Kit	Qiagen, Hilden
Thrombin CleanCleave Kit	Sigma-Aldrich Chemie, Buchs
Trizol kit	Invitrogen, Karlsruhe
S Tag Thrombin Purification Kit	Novagen, Nottingham

2.1.4 Radiochemicals

(α-32P)-dCTP, 3,000Ci/mmol Amersham Pharmacia, Freiburg
specific activity: 110Tbq/mmol
activity per volume: 370 Mbq/ m

2.1.5 Enzymes

Lysozyme	Sigma-Aldrich Chemie, Buchs
Proteinase K	Boehringer, Mannheim
Restriction Endonucleases	MBI Fermentas, St. Leon-Rot,
	New England Biolabs, Schwalbach
RNase A/T1 Mix	MBI Fermentas, St. Leon-Rot
T4-DNA-Ligase	MBI Fermentas, St. Leon-Rot
Taq-Polymerase	Qiagen, Hilden
Thrombin	Sigma-Aldrich Chemie, Buchs

2.1.6 Filters and Membranes

3MM Filter-Paper	Whatman Paper, Maidstone
Hyperfilm MP X-ray detection film	Amersham Pharmacia, Freiburg
Hybond-C extra Nitrocellulose membrane	Amersham Pharmacia, Freiburg
Hybond-N+ Nylon membrane	Amersham Pharmacia, Freiburg

2.1.7 Marker

Page Ruler Prestained Protein Ladder	Fermentas, St. Leon-Rot
RiboRuler RNA Ladders, High Range	Fermentas, St. Leon-Rot
RiboRuler RNA Ladders, Low Range	Fermentas, St. Leon-Rot

2.1.8 Antibodies

The following antibodies were used in this study (tab. 3).

Tab. 3: antibodies

Antibody	Source	Titer	Reference
Anti- mdnB	Rabbit, polyclonal	1: 5,000	
Anti- Poly- Histidin	Mouse, monoclonal	1:10,000	Sigma, Sparks
Anti-Mouse IgG Horseradish Peroxidase conjugate	Sheep	1:10,000	Amersham Pharmacia, Freiburg
Anti-Rabbit IgG FITC conjugate	Goat	1:100	Sigma, Sparks
Anti-Rabbit IgG Horseradish Peroxidase conjugate	Goat	1:10,000	Sigma, Sparks

2.1.9 Nucleic Acids

2.1.9.1 Plasmids and Fosmids

pDrive	Qiagen, Hilden
pet15b	Novagen, Nottingham
pACYC184	MBI Fermentas, St. Leon-Rot
pCC1Fos	Epicentre Technologies, Madison
pACYCDuet-1	Novagen, Nottingham

2.1.9.2 Primers

The primers used in this study are listed in the table below (tab. 4).

Tab. 4: Primers

Primer	Sequence (5´- 3´)	Application
degA fw	TTYGGNACYGAAGCNGNGG	Identification of *mca* in *M. aeruginosa* NIES298
degA rv	AGAAGACCAAGAACGAACTTCGCC	
mcaD Sonde fw	GTCTAGCTCATCGGCTACGG	northern blot probe preparation
mcaD Sonde rv	CAGGGTTCATCTCCCTGAAA	
mcaE 7806 fw	CGGGTTAACAAAGCAAACAA	northern blot probe preparation
mcaE 7806 fw	TGCAGCCAGTGATAGATGCT	
mcaA Sonde fw	GTGGAACCAGTTTTGCCACT	northern blot probe preparation
mcaA Sonde rv	TCCCGAAGTCATAACCAAGG	
mcaE NIES298 fw	CGGGTTAATAAAGCAAACAACA	northern blot probe preparation
mcaE NIES298 rv	TCCTACGCTTCGTCACCATC	
mcaE500 Fw	CCAACATCCCCTTTTAAGTTTTT	Screening of *mcaE* genes in various *Microcystis* strains
mcaE500 Rv	AGATAGGCTAATCAGTCGGATAGA	
all7011 fw	GAAGGTTTGCAATTTTGTCCA	northern blot probe preparation
all7011 rv	CGCCAACGGGATTAATTTCT	
mdnB fw	TTGGCTGGTTTTTGGGATAG	Screening of *mdnB* genes in various *Microcystis* strains
mdnB rv	CGATCGCATTGGAAATAGGT	
mdnB Express fw	TAAACTCAGATCACTTGAGTAATTCAGCACTTTT	Construction of the minimal ABCD cassette
mdnB Express rv	TGAAAGCACTGGAAAAACTGG	
mdnA Express fw	ACGCGTTAAGTAGTTGTGCAGCTATCAGT	Construction of the minimal ABCD cassette
mdnA Express rv	GGATCCTTTAATACTCTTCCCAGTCAGAAGG	
mdnB pET Exp fw	CATATGAAAGAATCGCCTAAAGTTGTTTTATTG	Overexpression of MdnB
mdnB pET Exp rv	GGATCCTAGCATACTAAAAAATCAGCGATCGCA	
mdnC pET Exp fw	CATATGATCTTTACTCAGGCGGTCAAAAAG	Overexpression of MdnC
mdnC pET Exp2 fw	CATATGACCGTTTTAATTGTTACTTTTAGCCACG	
mdnA pET Exp fw	CATATGGCATATCCCAACGATCAACAAGGT	Overexpression of MdnA
mdnA pET Exp fw	GGATCCTTAATACTCTTCCCAGTCAGAAGGGT	
mdnA fw multi	TCACTCGAAATTACCAGAGGAA	Screening of *mdnA* genes in various *Microcystis* strains
mdnA rv multi	CGGTGTAATCAAGAAAAGTGCT	

2.2 Methods

2.2.1 Cultivation of bacteria

2.2.1.1 Cultivation of cyanobacteria

The cyanobacteria were cultivated in BG-11 (Rippka et al., 1979) under continuous light of approximately 30 µmol photons $m^{-2}s^{-1}$ under continuous shaking with 40 rpm at 23°C. For the light experiment the cells were grown at 18 µmol photons $m^{-2}s^{-1}$ until they reached the required cell density and then exposed for two hours to different light intensities: 0 µmol photons $m^{-2}s^{-1}$ (dark, D); 18 µmol photons $m^{-2}s^{-1}$ (low light, L); 68 µmol photons $m^{-2}s^{-1}$ (high light, H) and 180 µmol photons $m^{-2}s^{-1}$ (very high light, VH). Light intensities were measured using a Li-Cor LI250 light meter (Walz, Effeltrich).

2.2.1.2 Cultivation of Escherichia coli

Escherichia coli cells were cultivated under standard conditions either in liquid LB medium or on LB agar in petri dishes (Sambrook et al., 1989). The *Escherichia coli* strain used for the fosmid library construction was cultivated according to the suggestions of the manufacturer (Epicentre Technologies, Madison). Cultures for preparation of plasmid vector DNA were incubated in 3 - 4 ml liquid LB medium at 37°C and shaking at 220 rpm. Corresponding to resistance markers the respective antibiotics were added in the appropriate concentrations.

2.2.2 Molecular biological techniques

2.2.2.1 Preparation of genomic DNA from Microcystis aeruginosa

Genomic DNA was prepared from *M. aeruginosa* NIES298 and MRC as reported previously (Franche, Damerval, 1988). Cells were harvested by centrifugation and the pellet was washed twice in TE-buffer (10 mM Tris-HCl, 1 mM EDTA, pH 8.0). The pellet was resuspended in 500 µl TE and lysozyme to a final concentration of 1 mg ml^{-1}. After incubation at 37°C for 1 h, EDTA, SDS and Proteinase K were added to final concentrations of 0.05 M, 2% and 50 µg ml^{-1}, respectively, followed by a further incubation at 50°C for 1 h. The mixture was extracted twice with phenole/chloroform/isoamylalcohol (25:24:1) and once with chloroform/isoamylalcohol (24:1). The final supernatant was precipitated with isopropanol (0,7 volume), washed with 70% ethanol and air-dried after centrifugation. The DNA was resuspended in water. To remove RNA from the extract, 1 µl of an RNase A/T1 Mix was added and samples were incubated at 37°C for 1 h.

2.2.2.2 Preparation of plasmid DNA from Escherichia coli

E. coli plasmid DNA was isolated either by standard procedure of the alkaline lysis (Sambrook *et al.*, 1989), or with the Plasmid MiniPrep kit (Qiagen, Hilden) according to manufacturer instructions.

2.2.2.3 Quantification of nucleic acids

Nucleic acid concentrations were determined using the NanoDrop ND100 Spectrophotometer (Peqlab Biotechnology GmbH, Erlangen) following the manufacturer instructions.

2.2.2.4 Digestion of DNA with restriction endonucleases

DNA molecules were digested with restriction endonuleases from either Fermentas (St. Leon-Rot) or New England Biolabs (Schwalbach) following the companies' protocols. Commonly, a reaction volume of 10 to 20 µl was chosen and DNA was incubated for 1 h at 37°C.

2.2.2.5 Polymerase chain reaction

DNA fragments were amplified by PCR using the Taq DNA-Polymerase System (Qiagen, Hilden). The web-based software "Primer3" (http://frodo.wi.mit.edu/cgi-bin/primer3/primer3_www.cgi) was used to design primers. Standard reactions involved initial denaturation at 95°C for 3 min, followed by 35 cycles of 20 s at 95°C for denaturation, 20 s at the respective annealing temperature (according to the melting temperature of the used primer pair) and 30 s - 15 min at 72°C (depending on the product size: 1 min per kilobase). An additional elongation step at 72°C for 10 min completed the reaction.

2.2.2.6 Agarose gel electrophoresis of DNA

DNA fragments were separated by agarose gel electrophoresis (Sambrook et al., 1989). Depending on the size of the fragments, 0.8 g- 1.2 g agarose (Biozym Diagnostik, Hameln) was melted in 100 ml TAE buffer (40 mM Tris, 20 mM acetic acid, 1 mM EDTA). Ethidium bromide was added to the gel in final concentration of 0.05 mg ml^{-1}. Samples were mixed with loading dye (50 % Ficoll; 1 mM EDTA, pH 8.0; 0.05 % (w/v) Bromophenole blue; 0.05 % (w/v) Xylene cyanol) and loaded into the agarose gel slots next to an aliquot from a *PstI* digest of phage λ DNA as size marker. Nucleic acids were visualised via a UV transilluminator (Gel doc XR System; BioRad, München) using the Quantity One software (BioRad, München).

2.2.2.7 Elution of DNA fragments from agarose gels

DNA fragments were eluted from agarose gels using the Jetsorb "Gel Extraction Kit" (Genomed, Löhne) according to the manufacturer's manual.

2.2.2.8 Purification of DNA

To purify DNA the Qiaquick PCR Purification kit (Qiagen, Hilden) was used as instructed by the manufacturer.

2.2.2.9 Ligation of linear DNA fragments into plasmid vectors

Purified PCR products were directly ligated in the pDrive vector using the PCR cloning kit (Qiagen, Hilden) as recommended by the manual. For heterologous expression of proteins, PCR products of the respective ORFs with introduced restriction sites of *NdeI* and *BamHI* (MBI Fermentas, St. Leon-Rot) were obtained and cloned in the pDrive vector as described. The fragment was cut out using the introduced restriction sites and ligated into the *NdeI* and *BamHI* digested pET15b vector (Novagen, Nottingham) using the T4 DNA ligase (MBI Fermentas, St. Leon-Rot).

2.2.2.10 DNA Sequencing and Sequence Analysis

DNA sequencing was performed by SMB GbR (Berlin). The fragments were assembled using the software package Vector NTI (Invitrogen, Karlsruhe). DNA sequences analysed via the NCBI (National Institute for Biotechnology Information, Bethesda, MD) BLAST server (http://www.ncbi.nlm.nih.gov/blast/).

2.2.2.11 Construction of a fosmid library and screening

DNA fragments of approximately 30 to 40kb were directly ligated to the pCC1FOS vector (Epicentre Technologies, USA) following manufacturer instructions. Screening of the library was performed by colony hybridisation using standard conditions (Sambrook *et al.*, 1989). Colony hybridisation was performed in hybridisation buffer containing 50% formamide at 42°C according to standard.

2.2.2.12 Transformation of Escherichia coli

Cells of *E. coli* were transformed utilising the $CaCl_2$-chemically induced competence (Sambrook *et al.*, 1989). An aliquot of 200 µl of the competent cells was mixed with the respective plasmid and kept on ice for 30 min. Heat shock was performed for 90 s at 42°C in a water bath. After addition of 500 µl SOC medium the samples were incubated at 37°C for 1 h and subsequently spread on LB agar plates containing the appropriate antibiotic. In case of blue-white selection of positive clones, 40 µl X-Gal solution (20 mg/ml in DMF) and 40 µl of IPTG solution (0.1 M) were added to the agar. The plates were incubated at 37°C overnight.

2.2.2.13 Radioactive labelling of DNA probes

Specific DNA fragments were radioactively labelled using the HexaLabel kit (Fermentas, St. Leon-Rot). Instructions provided by the manufacturer were followed, the radioactively labelled nucleotide used was α-32P-dCTP (Amersham Pharmacia, Freiburg) with an activity of 50 µCi in a volume of 5 µl.

2.2.2.14 Preparation of total RNA from Microcystis aeruginosa

Cells were harvested by centrifugation at 4°C and homogenised in liquid nitrogen using a pestle and mortar. RNAs were isolated using the Trizol kit (Invitrogen, Karlsruhe) according to the manufacturer's instructions.

2.2.2.15 Agarose gel electrophoresis of RNA

RNA was separated and visualised via formaldehyde gel electrophoresis (Sambrook *et al.*, 1989). 1,5 g agarose (Biozym Diagnostik, Hameln) was melted in 76 ml Milli-Q prior to the addition of 10 ml 10x MEN (200 mM MOPS, 50 mM NaAcetate, 10 mM EDTA, pH 7.0) and 4 ml formaldehyde solution. RNA samples were mixed with one volume of 2x

RNA Loading Dye (MBI Fermentas, St. Leon-Rot) and incubated at 65°C for 10 min. The samples were loaded into the gel slots next to an RNA high range or RNA low range marker (Fermentas, St. Leon-Rot) and separated by electrophoresis in 1x MEN at constant voltage of 60 V.

2.2.2.16 Northern Blot analysis

RNA separated by gel electrophoresis was immobilised on Hybond N+ Nylon membranes (Amersham Pharmacia, Freiburg) via capillary transfer (Sambrook *et al.*, 1989) with 6x SSC buffer (0.09 M NaCitrate, 0.9 M NaCl, pH 7.0) for at least 12 h. The membranes were prehybridised with the Northern buffer (50 % (v/v) deionised formamide, 0.25 M NaCl, 7 % (w/v) SDS 0.125 M Na_2HPO_4; pH 7.2) for 1h in a hybridisation oven at 42°C. The DNA probe was denatured at 95°C for 5 min and subsequently added to the hybridisation tube. The hybridisation continued over night at 42°C. The membranes were washed successively for 15 min in Solution 1 (0.5% SDS (w/v); 2x SSC) at 50°C and twice in Solution 2 (0.1% SDS (w/v); 0.1x SSC) at 60°C. Signals were detected using the BioRad Imaging Screen-K screens and the Personal Molecular Imager FX (BioRad, München).

2.2.3 Protein biochemical methods

2.2.3.1 Preparation of proteins from Microcystis aeruginosa

Cells were harvested by centrifugation for 10 min at 4,000 g. Pellets were resuspended in 250 ml buffer A (500 mM Tris-HCl, 50 mM EDTA; pH 7.2) and one volume of glassbeads (d 0,11 and 0,18 in a 1:1 ratio). The cells were lysed by three cycles of freezing in liquid nitrogen and thawing, alternating with 10 min of treatment in a Mixer Mill MM2 (Retsch, Haan). After 10 min of centrifugation at 13,000 rpm the supernatant contained the soluble proteins. The membrane proteins were extracted by resuspending the pellet in 250 µl of buffer B (100 mM NaH_2PO_4, 10 mM Tris-HCl, 8 M Urea, pH 8.0) and repeating the treatment in the glass bead mill.

2.2.3.2 Heterologous expression of proteins

For the expression of recombinant proteins in *E. coli*, 300 ml LB medium containing 50 µg ml^{-1} ampicillin were inoculated with an overnight culture of the respective *E. coli* clone and grown at 37°C with vigorous shaking (220 rpm) until the OD$_{600}$ was 0.5 to 0.7. The expression was induced by addition of IPTG to a final concentration of 0.5 mM and the cells were grown for 3 h at 18°C or 37°C. The cells were harvested by centrifugation at 4,000 g for 10 min.

2.2.3.3 Purification of recombinant proteins from Escherichia coli

Recombinant proteins expressed from the pET15b vector (Novagen, Nottingham) were purified using the Ni-NTA Superflow (Qiagen, Hilden) according to manufacturer instructions (The QIA*expressionist*). For native purification of his-tagged proteins, cell pellets were resuspended in 10 ml lysis buffer (50 mM NaH$_2$PO$_4$, 300 mM NaCl, pH 8.0) and sonicated on ice for 1 min. The cellular debris was pelleted by centrifugation at 13,000 rpm for 10 min at 4°C. Afterwards, 1 ml Ni-NTA slurry (Qiagen, Hilden) and Imidazol to a final concentration of 30 mM were added to the supernatant and mixed gently by shaking on a rotary shaker at 4°C at least for 1 h. Subsequently the mixture was loaded into 1 ml - Polypropylene Columns (Qiagen, Hilden), washed twice with 4 ml wash buffer (lysis buffer containing 50 mM imidazol) and eluted four times with 0.5 ml elution) buffer (lysis buffer containing 250 mM imidazol). Protein purity was estimated by SDS-PAGE gel electrophoresis. To purify his-tagged proteins under denaturing conditions, the same procedure was performed though different buffers were used. Cellpellets were resuspended in buffer B (100 mM NaH$_2$PO$_4$, 10 mM Tris-HCl, 8 M urea; pH 8.0), the lysate-resin mixtures in Polypropylene Columns were washed twice with 4 ml buffer C (buffer B; pH 6.3) and eluted four times with buffer D (buffer B; pH 5.9) and four times with buffer E (buffer B; pH 4.5).

S-tagged proteins were purified using the S Tag Thrombin Purification Kit (Novagen, Nottingham) following the manufacturer instructions. Elution was performed with 0.5 ml 10x bind/wash buffer (200 mM Tris-HCl pH 7.5, 1.5 M NaCl, 1% Triton X-100) added with $MgCl_2$ to a final concentration of 3 M.

2.2.3.4 Determination of protein concentration

The concentration of proteins in various samples was determined using the Protein Assay (BioRad, München) according to manufacturer instructions. Absorbance was measured at 595 nm and the concentrations were calculated using standard curve generated by dilutions of BSA.

2.2.3.5 SDS-Polyacrylamide gel electrophoresis (SDS-Page)

Proteins were separated by the method of discontinuous gel electrophoresis (Laemmli, 1970) added with SDS to obtain denaturing conditions. The gels consisted of a separating gel containing 10-15% acrylamide (10-15% (v/v) acrylamide/bisacrylamide 37.5:1 (v/v), 375 mM Tris-HCl pH 8.8, 0.1% (w/v) SDS) depending on the protein size range that was to be examined and a stacking gel of 4% acrylamide (4% (v/v) acrylamide/bisacrylamide 37.5:1 (v/v), 125 mM Tris-HCl pH 6.8, 0.1% (w/v) SDS). Samples were mixed with 5x protein loading dye (250 mM Tris pH 6.8, 0.5% bromophenole blue, 10% (w/v) SDS, 50% (v/v) Glycerol, 500 mM 2-mercaptoethanol) and denaturated at 95°C for 5 min. The gels were run at a constant current of 25 mA per gel in the Mini Protean II system (BioRad, München). Gels were either stained in GelCode Blue Stain Reagent (Pierce, Rockford) or kept for further analyses.

2.2.3.6 Western Blot analysis

Proteins were immobilised on Hybond C-extra Nitrocellulose membranes (Amersham Pharmacia, Freiburg) for immunodetection. Blotting was performed using the Mini TransBlot Cell system (BioRad, München). Gels and accordingly sized membranes were equilibrated for 10 min in Western blot transfer buffer (15.6 mM Tris; 120 mM Glycine). Prior blotting "sandwich" was prepared following manufacturer's instructions. Blots were run at a constant current of 380 mA for 1 hour in Western Blot transfer buffer.

For immunodetection of specific proteins, membranes were blocked with 5% w/v milk powder in PBS-T (140 mM NaCl; 2.7 mM KCl; 8 mM Na_2HPO_4; 18 mM KH_2PO_4, pH 7.4; 0,01 % Tween 20) for 1 h and subsequently probed with the primary antibody in PBS-T for 1h. The membranes were washed three times with 25 ml of PBS-T followed by 1 h of incubation in secondary antibody solution. After three times of final washing in PBS-T, visualisation of band signals was performed using the SuperSignal West Pico Chemiluminescent Substrate kit (Pierce, Rockford) according to manufacturer's instructions in combination with X-ray films or the Lumi-Imager (Boehringer, Mannheim) for detection.

2.2.3.7 Dialysis of proteins

Dialysis tubings (SERVA Feinbiochemika, Heidelberg) with an exclusion size of 12 kDa were pretreated according to the manufacturer's recommendations and loaded with the samples. Dialysis was performed in 2 L of the buffer of choice at 4°C or room temperature under constant stirring over night. The buffer was changed three times.

2.2.3.8 Generation of an antibody against MdnB

To obtain a specific antibody against MdnB, his-tagged MdnB was purified under denaturing conditions. The purified protein was used to raise a polyclonal rabbit antibody (Pineda-Antibody-Service, Berlin). Serum samples were taken every 30 days and tested by immunoblotting with protein extracts from *M. aeruginosa* NIES298 and purified his-tagged MdnB.

2.2.3.9 In vitro assays with MdnB and MdnC

The purified his-tagged MdnB and MdnC were assayed for cyclisation activity with two different synthetic prepeptides (Genscript, Piscataway). Prepeptide 1 consist of 14 amino acids encoding Microviridin B (FGTTLKYPSDWEEY). Prepeptide 2 is identical in amino acid sequence with MdnA, the whole precursor protein for Microviridin B synthesis in *M. aeruginosa* NIES298 (MAYPNDQQGKALPFFARFLSVSKEESSIKSPSPEPTFGTTLKYPSDWEEY). Reaction mixtures contained 10 µg of the respective Prepeptide, 2,5 mM ATP and 1 -5 µg purified MdnB or MdnC in 500 µl of 3x MdnB/C buffer (600 mM Tris-HCL pH 8.0, 60 mM $MgCl_2$, 300 mM NaCl). Additionally, different buffer compositions were tested, varying the concentration of Tris –HCl from 100 to 1000 mM, of MgCl from 10 to 100 mM and of NaCl from 100 to 600 mM. PH values of 6.8, 7.5 and 8.6 were tested. Sometimes, GTP was used instead of ATP. The reaction was incubated at room temperature or 37° C for at least 16 h. Aliquots of 100 µl were taken after 10 min, 1 h, 5 h and 16 h and immediately concentrated in vacuo. Samples were analysed via HPLC or send to Dr. Keishi Ishida (Hans Knöll institute, Jena) for further chemical characterisations.

2.2.3.10 Heterologous expression of peptides from fosmids

A total volume of 100 ml of the *E. coli* clone was induced to high copy number according to the suggestions of the manufacturer (Epicentre Technologies, Madison). The cells were grown for five hours at 37°C and subsequently harvested by centrifugation at 4,000g for 10 min.

2.2.3.11 Extraction of peptides from Microcystis aeruginosa and Escherichia coli

The harvested cells were resuspended in deionised water and lysed by sonication. The cellular debris was removed by centrifugation. The resulting supernatant was loaded on a Sep-Pak cartridge (Waters Corporation, Eschborn), which then was washed with 5% methanol. The components of interest were eluted with 100% methanol and concentrated *in vacuo*. Died extracts were resuspended in 200µl 50% methanol, filtered (Acrodisc 4 mm Syringe Filter; 0,45 µm Nylon Membrane; PALL, East Hills) and subjected to reversed phase column HPLC.

2.2.3.12 HPLC analysis

The HPLC separation of cell-extract and supernatant samples was conducted on a Shimadzu HPLC unit comprising the system controller SCL-10AVP, the pump LC-10Ai, the autosampler SIL-10A, the fraction collector FRC-10A and the Photodiode-array-detector (PDA-detector) SPD-M-10AVP. Separation was carried out on a "SymmetryShield RP18" column (Waters) with a particle size of 3.5 µm, 4.6 mm inner diameter and 100 mm length and a precolumn (3.9 mm x 20 mm) with an identical sorbent. The following gradient system was used at a flow rate of 1.0 ml min^{-1}: 1 min loading (20% buffer B), linear gradient up to 80% buffer B in 30 min, followed by a linear gradient to 100% buffer B in 1 min, and then holding 100% buffer B for 3 min (buffer A, 0.1% TFA in H_2O; buffer B, acetonitrile, 0.1% TFA).

2.2.3.13 Chemical Drawings

Chemical drawings were produced using the software ChemDraw Ultra 12.0 (CambridgSoft, Cambridge).

2.2.4 Immunofluorescence Microscopy

2.2.4.1 Fixation and permeabilisation of cyanobacteria

Cyanobacteria were prepared for Immunofluorescence Microscopy as reported previously (Guljamow et al., 2007). 30 ml of cell culture were harvested by centrifugation at 4000 g for 10 min, washed once in PBS (140 mM NaCl, 2.7 mM KCl, 8 mM Na_2HPO_4, 1.8 mM KH_2PO_4, pH 7.4) and resuspended in 200 µl PBS. Cells were fixed in 3.7% (v/v) formaldehyde in PBS for 1 h on ice. After three times of washing in PBS, cells were resuspended in GTE buffer (50 mM glucose; 20 mM Tris-HCl pH 7.5; 10 mM EDTA, pH 8.0) and permeabilised by addition of freshly prepared lysozyme solution to a final concentration of 2 mg/ml. After 4 min lysozyme incubation, cells were spread-to-dry on poly- L-lysine coated glass slides. Glass slides with immobilised cells were dipped into methanol for 5 min at -20°C and subsequently into -20°C acetone for 30 sec.

2.2.4.2 Immunostaining of fixed cyanobacterial cells

Fixed cells were blocked in 2% (w/v) BSA in PBS for 15 min and incubated with the primary antibody at a dilution of 1:500 in 2 % BSA (w/v in PBS) at RT for 1 h in a humid chamber. Afterwards, slides were washed twice in PBS for 10 min and subsequently incubated with a FITC-labelled secondary antibody at a dilution of 1:100 in 2% BSA for 1 hour in a humid chamber. Hoechst DNA dye (in a final concentration of 0.05 µg/ml; Serva, Heidelberg) was added to the second of two PBS washes. Cells were mounted in a drop of 4% (v/v) n-propylgallate dissolved in 87% (v/v) glycerol and stored for up to 4 weeks at -20°C.

2.2.4.3 Fluorescence microscopy: image acquisition and processing

Sample observation, image acquisition and processing were carried out using the DeltaVision spectris system (Applied Precision) with the pre-installed default softWorx software package. Three sets of excitation and emission filters were used for visualisation: the "RD-TR-PE" filter pair (555 nm/617 nm) to visualise red/orange autofluorescence of cyanobacteria, the "FITC" filter pair (490 nm/528 nm) to visualise FITC-coupled green immunostaining and the "DAPI" excitation/emission filter pair (wavelengths of 360 nm and 457 nm, respectively) to visualise Hoechst-stained DNA, appearing as blue in the acquired images. Acquired raw images were deconvolved by iterative constrained deconvolution to enhance image quality and contrast using the algorithms implemented in the softWorx software package.

2.2.5 Phylogenetic analysis

2.2.5.1 CLANS (cluster analysis of sequences)

For rapid classification and visualisation of peptide clusters the program CLANS (http://bioinfoserver.rsbs.anu.edu.au/programs/clans) was used (Frickey, Lupas, 2004). Peptide sequences were provided in a FASTA file containing unaligned amino acid sequences. E-values were specified as 0.1. The procedure was run for approximately 200 rounds.

2.2.5.2 Retrieval and alignment of sequence data

Sequence data was retrieved by searching available databases through the NCBI internet gateway at http://www.ncbi.nlm.nih.gov. Amino acid sequences were aligned using the multiple alignment algorithms implemented in the ClustalX program version 1.83 (Thompson et al., 1997a; Thompson et al., 1997b) with the BLOSUM 62 protein weight matrix.

Amino acid based alignments were manually edited to remove large gaps and ambiguously aligned regions using the BioEdit v7.0.5.3 program (Hall, 1999). Both programs were also used to convert alignment files into formats required as input by the phylogenetic programs employed in further analyses.

2.2.5.3 Reconstruction of phylogenetic trees

For the Bayesian method (Larget, Simon, 1999) the software MrBayes v3.1.2 (Huelsenbeck, Ronquist, 2001) was utilised. This program required the setting of a prior probability distribution for the parameters of the likelihood function set, here, the Jones-Taylor-Thornton matrix (Jones et al., 1992) was chosen to model substitution rates. The algorithm was started instructing it to calculate 500,000 generations with 4 chains, while sampling every 100th generation to create a tree whose branch length was to be recorded by the program. At conclusion of each run, after a reasonable burn-in was chosen (usually 50% of all tree data), trees were summed up in a consensus tree, which was to include all compatible groups.

Maximum likelihood analyses were performed utilizing the quartet-puzzling method implemented in the TREE-PUZZLE v5.2 program (Schmidt et al., 2002a; Schmidt et al., 2002b; Strimmer, von Haeseler, 1996; Strimmer, vonHaeseler, 1996). The program used the aligned sequence file in .phy format as input and was subsequently instructed to calculate a quartet-puzzling tree in 25,000 puzzling steps while approximating missing parameters using the NJ method (Saitou, Nei, 1987). Furthermore, JTT was chosen as the model for substitution rates while γ distribution parameters of rate heterogeneity over sites were estimated from the data set with the aid of 8 distinct categories.

3 Results

3.1 Microcyclamides

3.1.1 Identification of microcyclamide biosynthesis genes in *Microcystis aeruginosa* NIES298

Microcyclamide is a cytotoxic cyclic hexapeptide isolated from the cultured cyanobacterium *Microcystis aeruginosa* NIES298 (Ishida et al., 2000). Containing two thiazole and one methyloxazole moiety it belongs to the cyanobacterial peptide family of cyclamides (Welker, von Döhren, 2006). Since for structurally related patellamides a ribosomal assembly line was shown (Schmidt et al., 2005), we proposed a similar biosynthetic pathway for microcyclamide. Two patellamide-like gene clusters were known by the time of this study - one responsible for the biosynthesis of patellamides in *Prochloron didemni* (Schmidt et al., 2005) and the other for trichamide in *Trichodesmium erythraeum* (Sudek et al., 2006). These studies revealed that although the precursor proteins are highly variable, the processing enzymes are quite conserved exhibiting 45 to 60% identical amino acids to their homologues.

To detect orthologous genes in *M aeruginosa* NIES298, we therefore designed a set of degenerated primers based on the *patA* sequence from *P. didemni* and the corresponding *triH* from *T. erythraeum* (fig. 12).

```
                    Primer degA fw
   patA      ...TACGACTTCGGCACCGAAGCTCGCCGCGACACCTTCAAGCAACTGATGCCGCCCTTTGAC
   triH      ...TATGACTTTGGAACTGAAGCACGGCGGGACTCATTTAAACAGTTGATGCCAGCAGTTACT     1380
              ** ***** ** ** *****  ** ** *** * ** ** **  *******  *  **

   patA      ...GTGAGCGATCTGGTTCCAGTGACTATGGGCGAAGTTCGTTCTTGGTCTTCTTCTTACTAA
   triH      ...GTTAGCGATTTCATTCCGGTAACTTTGGGCGAAGTTCGTTCTTGGTCTTCTCCTTATTAG     2127
              ** ******  *  **** ** *** ******************** ***** **** **
                                            Primer degA rv
```

Fig. 12: Nucleotide sequence alignment of *patA* in *P. didemni* and *triH* in *T. erythraeum*. The asterisks indicate identical bases at the corresponding positions. Green arrows show where the degenerated Primers were deduced from.

PCR amplification using genomic DNA of *M. aeruginosa* NIES298 yielded a product of the expected size of approximately 800bp (data not shown). Subsequent sequencing of the PCR product revealed 79% and 68% identity in nucleotides to the *patA* and *triH* gene, respectively. The PCR product was used to obtain a radioactively labelled probe, which then was hybridised with colonies on nylon membranes of a fosmid library constructed from genomic DNA from *M. aeruginosa* NIES298. Sequencing of the positive clone led to the discovery of a putative patellamide-like assembly line comprising about 13 kb. Flanked by two genes encoding a putative transposase and a restriction-modification enzyme, the cluster contains homologues of all seven genes found in the *pat* cluster and two additional ORFs. The putative precursor directly encodes the microcyclamide amino acid sequence HCATIC (Fig. 13B). Thus, the cluster could unambiguously be assigned to microcyclamide biosynthesis and the genes were designated *mca* (fig. 13).

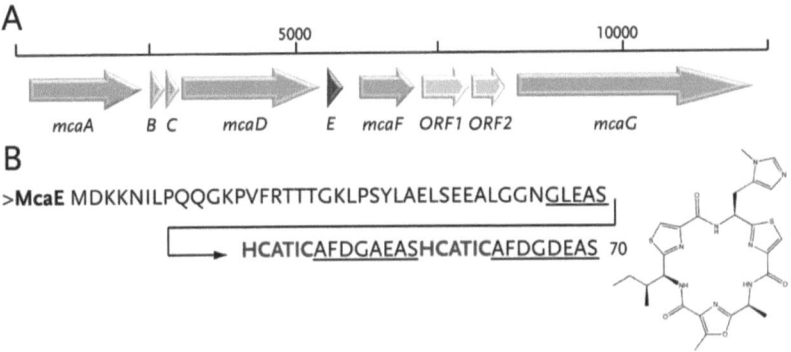

Fig. 13 Microcyclamide biosynthesis genes in *M. aeruginosa* NIES298.
(A) Schematic representation of the *mca* biosynthetic gene cluster. Genes with similarity to patellamide biosynthesis genes in *P. didemni* are in gray. The precursor protein is highlighted in black. The light grey arrows represent ORFs that could not directly be assigned to microcyclamide biosynthesis. (B) McaE sequence of NIES298. The sequences encoding microcyclamide are in green. The proposed start and stop cyclisation sequences are underlined. The structure of microcyclamide is shown on the right.

3.1.2 Comparative analysis of the *mca* genes

Within the time of this study more *pat*-like gene clusters were discovered and published. BLAST analysis of the encoded proteins of the *mca* gene cluster showed hits for patellamide-like biosynthesis proteins in a variety of cyanobacteria. The top scoring BLAST hits for five of the *mca* genes were annotated as hypothetical proteins in the genome of the mat-forming cyanobacterium *Lyngbya* sp. PCC8106 (tab. 5). So far no corresponding metabolite could be assigned for the gene cluster in *Lyngbya*. Other BLAST hits showed proteins responsible for the biosynthesis of tenuecyclamide in *Nostoc spongiaeforme var. tenue* and patellins in an uncultured *Prochloron* strain. Like the *mca* genes both of these gene clusters are syntenic to the patellamide biosynthesis cluster (Donia et al., 2008). Since the *pat* genes have been the first known and best characterised of these assembly lines, we used this cluster as a basis for our comparisons.

Tab. 5: Deduced functions of ORFs in the microcyclamide biosynthesis cluster in *M. aeruginosa* NIES298

Protein	Length (aa)	Protein-ID	Deduced Function	Sequence Similarity	Identity/ Similarity (aa length)
McaA	657	CAO82081	Subtilisin-like protease	PatA; *Prochloron didemni*	68%/78% (702)
McaB	83	CAO82082	Unknown	PatB, *Prochloron didemni*	72%/79% (83)
McaC	80	CAO82083	Unknown	PatC; *Prochloron didemni*	53%/67% (70)
McaD	776	CAO82084	Adenylation/ hetero-cyclisation	PatD; *Prochloron didemni*	77%/86% (785)
McaE	74	CAO82085	Microcyclamide precursor protein	Patellamide precursor protein; *Prochloron didemni*.	68%/78% (69)
McaF	321	CAO82086	Unknown	PatF; *Prochloron didemni*	52%/69% (311)
McaG	1351	CAO82089	Thiazoline oxidase/ subtilisin-like protease	PatG; *Prochloron didemni*	71%/81% (729)
ORF1	267	CAO82087	Unknown	Hypothetical; *Lyngbya sp.* PCC8106	90%/94% (267)
ORF2	115	CAO82088	Unknown	PatG; *Prochloron didemni*	54%/70% (37)

3.1.2.1 The posttranslational machinery involved in microcyclamide biosynthesis

The *mcaA* gene encodes a multidomain enzyme with 68% identity to PatA from *P. didemni*. Whereas no conserved motif was detected in the C-terminal part, the N-terminal region is similar to subtilisin-like proteases, which usually are involved in the posttranslational tailoring of peptide pheromones in gram-positive bacteria (van der Meer et al., 1993). Since PatA has been shown to catalyse the proteolytic cleavage of the N-terminal recognition sequence from the precursor peptide (Lee et al., 2009), McaA is expected to play a similar role in microcyclamide biosynthesis.

McaB has 72% identity to PatB. According to BLASTp analysis all proteins homologous to McaB/PatB are connected to the biosynthesis of patellamide-like peptides. Therefore, no predicted function could be assigned to these gene products. PatB has been shown to be dispensible for patellamide biosynthesis (Donia et al., 2006). However, the presence of homologues in all patellamide-like gene clusters and the high identity over their entire length, suggest a characteristic role in peptide biosynthesis.

The deduced gene product of *mcaC* is 53% identical to PatC, which is not essential for heterologous expression of patellamides (Donia et al., 2006). Like McaB, McaC shows no significant similarity to any characterised protein in the database.

McaD contains two domains. The entire McaD peptide sequence is 77% identical to PatD and similar to several proteins involved in the biosynthesis of different metabolites. It shows low sequence similarity at the N-terminus to the bacteriocin biosynthesis cyclodehydratase - SagC family of ezymes, including McbB from *E. coli* (Zamble et al., 2000) and SagC from *Streptococcus pyogenes* (Lee et al., 2008). The C terminus is similar to the SagD family of enzymes. The McbBCD complex in *E. coli* and the SagBCD complex in *Streptococcus pyogenes* have been shown to be responsible for the heterocyclisation of cysteine residues in microcin and streptolysin S biosynthesis (Lee et al., 2008). Therefore, patD is proposed to be involved in heterocyclisation of cysteine and serine or threonine into thiazoline and oxazoline rings.

Furthermore, the BLAST analysis revealed similarities to gra-orf12, which belongs to the biosynthesis cluster of granaticin, a benzoisochromaquinone-type antibiotic produced by *Streptomyces violaceoruber* (Ichinose et al., 1998) and GodD, a protein from the goadsporin gene cluster in *Streptomyces* sp. TP-A0584 (Onaka et al., 2005). Uncharacterised proteins similar to McaD are detectable in taxa such as Cyanobacteria, Myxobacteria and other Proteobacteria, suggesting a widespread occurrence of these enzymes.

McaF shares 52% identity with PatA from *P. didemni*. The protein was shown to be essential in patellamide biosynthesis (Donia et al., 2006). Although it is not encoded in the trichamide gene cluster (Sudek et al., 2006), two copies are found in the *tru* pathway for trunkamide and patellin biosynthesis in an uncultured *Prochloron* strain (Donia et al., 2006). As the protein does not contain any conserved domain motif, no functional role in microcyclamide biosynthesis could be deduced.

A PatG homologue with 71% identity is encoded at the 3´ end of the microcyclamide cluster. The *N*-terminal region of McaG revealed characteristic features of NAD(P)H oxidoreductases and is distantly related to McbC from microcin B17 biosynthesis (Gehring et al., 1998). It is therefore predicted to oxidise thiazoline rings into the thiazole oxidative state. The C terminal half of McaG contains a subtilisin-like protease, whose homologous region in *P. didemni* has been shown to be responsible for cleavage and cyclisation of the precursor protein in patellamide biosynthesis (Lee et al., 2009).

Two ORFs present in the microcyclamide cluster are not part of the trichamide and patellamide gene clusters. The first one encodes a protein of 267 amino acids that shows no significant similarity to any protein in the database except a hypothetical protein encoded in the orthologous peptide gene cluster from *Lyngbya*. The second ORF encodes a small protein of 115 amino acids with partial similarity to McaG, suggesting that this gene could represent a pseudogene.

3.1.2.2 The Precursor: mcaE

The alignment of known *pat*-like precursor proteins in figure 14 shows that the leader peptides are quite conserved. In particular, the last five amino acids (AELSEEAL) of the leader sequences are identical in four of five precursor peptides. The proposed cyclisation signals surrounding the microcyclamide-coding region (GAEAS. . .AFD) resemble the corresponding parts of the other precursors, except for the trichamide-coding region. However, the copy numbers of encoded peptides differ. Whereas the patellamide precursor and the TruE1 protein in *Prochloron* strains encode two different peptides, two copies of the same microcyclamide are encoded in *M. aeruginosa* NIES298. TenE encodes two copies of the primary amino acid sequence of each of the hexapeptides tenuecyclamide A and C, for a total of four copies. The trichamide precursor is proposed to encode just one peptide.

```
PatE    MNKKNILPQQGQPVIRLTAGQLSSQLAELSEEALGDAGLEASVT---------ACIT--FCAYDGVEPSITVCISVCAYDGE-----------
TruE1   MNKKNILPQLGQPVIRLTAGQLSSQLAELSEEALG--GVDASTL---------PVPT--LCSYDGVDASTVP--TLCSYDD-----------
TenE    MDKKNILPQQGKPVIRITTGKLPSFLAELSEEALGDAGVGAASATGCMCAYDGAGASATGCMCAYDGAGASATA--CACAYDGAGASATACACAYE
McaE    MDKKNILPQQGKPVFRTTTGKLPSYLAELSEEALGGNGLEASHC---------AT--ICAFDGAEASHCA--TICAFDGDEA----------
TriG    MGKKNIQPNSSQPVFRSLVAR--PALEELREENLTEGNQGHGPL---------ANGPGPSGDGLHPRLCS----CSYDGDDE----------
        *.****.*:..:**:*  ..:  . * ** ** .  .             : **  .        *::*.
```

Fig. 14 Alignment of different *pat*-like precursor proteins in cyanobacteria.
The leader peptides are shaded in grey, the proposed recognition sequences are shown in bold letters. The peptide coding sequences are shown in green, the double glycine motive in red letters.

A unique feature of the microcyclamide precursor peptide is the presence of a double glycine motif at the C-terminal part of the leader sequence, suggesting a possible transport through the inner cell membrane, since this motif is known to play a role in peptide translocation in different groups of bacteria (Michiels *et al.*, 2001). However, analysis of culture supernatants of *M. aeruginosa* NIES298 showed no microcyclamide-like peptides.

3.1.3 Heterologous expression of microcyclamide

Heterologous expression of secondary metabolite gene clusters is an important precondition for biotechnological applications, in particular when the biosynthesis genes are hosted in nonculturable bacteria such as *Prochloron* or in slow-growing bacteria such as planktonic cyanobacteria or myxobacteria (Wenzel, Müller, 2005). Since patellamides were successfully expressed from fosmids (Long et al., 2005; Schmidt et al., 2005), cell pellets and culture supernatants of *E. coli* cells harbouring the *mca* fosmid in high copy numbers were analysed using HPLC to test for direct expression of microcyclamides. However, no microcyclamide-like structures were detectable under different conditions and using different growth media, although *mca* transcripts were clearly visible in Northern Blot analysis (data not shown).

3.1.4 Transcription of the *mca* genes in *M. aeruginosa* NIES298

Several of the peptide bacteriocins of Gram-positive bacteria like subtilin, nisin and other lantibiotics are primarily produced in the late growth phase and act as autoinducers of their own biosynthetic genes and other target genes (Kleerebezem et al., 1997). To elucidate the expression profile of microcyclamide genes in *M. aeruginosa* NIES298, we performed DNA hybridisations with RNA prepared from cells incubated under different light intensities and collected at different growth phases (fig. 15).

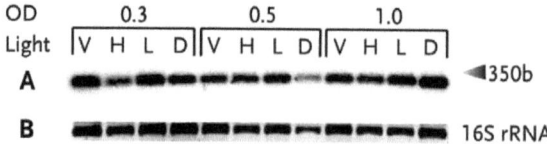

Fig. 15 Transcription analysis of the *mcaE* precursor RNA from *M. aeruginosa* NIES298.
Total RNA isolated from cells incubated under different light intensities and collected at different cell densities (OD, optical density at 750 nm). D, dark (0 µmol photons m^{-2} s^{-1}); L, low light (16 µmol photons m^{-2} s^{-1}); H, high light (68 µmol photons m^{-2} s^{-1}); V, very high light (180 µmol photons m^{-2} s^{-1}). (A) Autoradiogram of DNA-RNA hybridisation with an mcaE probe. (B) Agarose gel picture of the 16S rRNA stained with ethidium bromide under UV light. b, bases.

Hybridisation with different *mca* probes revealed a more or less constitutive transcription, independent from the growth phase. Similarly, different light intensities had no significant effect on the transcript abundance of the microcyclamide genes. Depending on the probe used different transcript sizes were detected. Whereas hybridisation with a *mcaE* probe revealed a transcript of about 350 bases, probes for *mcaA* and *mcaD* detected comparatively weak signals with DNA fragments of higher sizes (data not shown). These data suggest an independent transcription of the precursor at a higher rate than the putative tailoring enzymes. Since the precursor is used as substrate, this observed status of transcript levels seems plausible

3.1.5 An orphan microcyclamide-like gene cluster in *M. aeruginosa* PCC7806

3.1.5.1 Identification of homologous genes in strain PCC7806

Although *M. aeruginosa* NIES 298 is the only strain of the genus *Microcystis* that was previously shown to produce a microcyclamide-like structure, the genetic potential to produce this family of cyclic peptides may be far more widespread than expected. In order to test this hypothesis, the genome database (AM778843 to AM778958) of a second strain, *M. aeruginosa* PCC7806, was searched for the presence of similar genes. BLASTp analysis revealed a gene cluster encoding proteins with more than 90% identity to the nine proteins encoded in NIES298 (fig. 16 and tab. 6), suggesting the capability of strain PCC7806 to produce microcyclamide-like peptides.

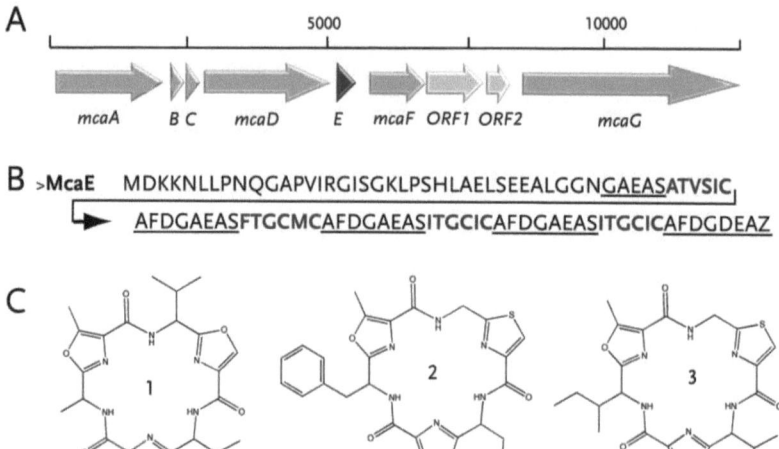

Fig. 16 Microcyclamide biosynthesis genes in *M. aeruginosa* PCC7806.
(A) Schematic representation of the *mca* biosynthetic gene cluster. Genes with similarity to patellamide biosynthesis genes in *P. didemni* are in gray. The precursor protein is highlighted in black. The light grey arrows represent ORFs that could not directly be assigned to microcyclamide biosynthesis. (B) McaE sequence of PCC7806. The sequences encoding microcyclamide are in green. The proposed start and stop cyclisation sequences are underlined. (C) Predicted microcyclamide structures in strain PCC7806.

Detailed analysis of the precursor peptide McaE in strain PCC7806 revealed an almost identical leader peptide and the same cyclisation signals as in *M. aeruginosa* NIES298. However, the peptide-coding sequences differ. Whereas the precursor in NIES298 contains two peptide-coding regions, the corresponding protein in strain PCC7806 encodes four (fig. 16B). Using microcyclamide from NIES298 as a model, three different hexapeptides containing thiazole and oxazole rings could be predicted (fig. 16C).

Tab. 6: Deduced functions of ORFs in the microcyclamide biosynthesis cluster in *M. aeruginosa* PCC7806

Protein	Legth (aa)	Protein-ID	Deduced Function	Sequence Similarity	Identity/ Similarity (aa length)
McaA	657	CAP64335	Subtilisin-like protease	PatA; Prochloron didemni	69%/78% (702)
McaB	83	CAP64336	Unknown	PatB, Prochloron didemni	69%/78% (83)
McaC	80	CAP64337	Unknown	PatC; Prochloron didemni	50%/65% (72)
McaD	776	CAP64338	Adenylation/ hetero-cyclisation	PatD; Prochloron didemni	77%/86% (785)
McaE	97	CAP64339	Microcyclamide precursor protein	Patellamide precursor protein; Prochloron didemni.	60%/74% (70)
McaF	321	CAP74572	Unknown	PatF; Prochloron didemni	54%/70% (312)
McaG	1313	CAP64342	Thiazoline oxidase/ subtilisin-like protease	PatG; Prochloron didemni	71%/81% (729)
ORF1	267	CAP64340	Unknown	Hypothetical; Lyngbya sp. PCC8106	91%/95% (267)
ORF2	115	CAP64341	Unknown	PatG; Prochloron didemni	54%/70% (37)

The accurate prediction of chemical structures produced by cryptic pathways can guide the structure-based detection and elucidation of the corresponding compounds (Gross, 2007). We therefore initiated a reanalysis of the metabolite spectrum from *M. aeruginosa* PCC7806 to search for the predicted microcyclamide-like compounds.

3.1.5.2 Isolation of new microcyclamides

To identify the cryptic microcyclamide-like peptides in *M. aeruginosa* PCC7806, we first attempted their heterologous expression with fosmids containing the *mca* gene cluster in *E. coli*. Since this strategy failed, we checked the transcription of the *mca* genes in *Microcystis aeruginosa* PCC7806. Similar transcription patterns as in strain NIES298 (data not shown) suggested a more or less constitutive production of the peptides. Using the predicted structures as guides, isolation and structure elucidation of two new microcyclamides, microcyclamide 7806A and 7806B (fig. 17), were performed in cooperation with Keishi Ishida and Christian Hertweck at the HKI in Jena (Ziemert et al., 2008).

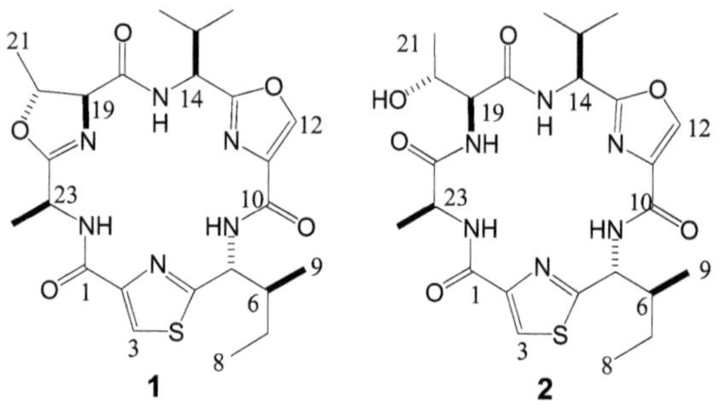

Fig. 17 Structures of microcyclamide 7806A (1) and 7806B (2)

The amino acid sequence of both peptides could be deduced as ATVSIC, corresponding to the first peptide encoding sequence of McaE in *M aeruginosa* PCC7806. The two microcyclamides were tested against HeLa cells. However, both peptides did not show any inhibitory activity. Microcyclamides 7806A and 7806B were also negative in standard antiproliferative, antibacterial, and antifungal assays (data not shown).

Further studies by C. Portmann and colleagues detected aerucyclamides A to D, corresponding to all three peptide encoding amino acid sequences of McaE (Portmann et al., 2008a; Portmann et al., 2008b). In contrast to the predicted metabolite, microcyclamide 7806A, 7806B and aerucyclamides are not fully oxidised. However, genomic analysis guided the structure elucidation of these new compounds, giving a new example of a successful genome mining approach in cyanobacteria.

3.1.6 Variability of microcyclamides in *Microcystis*

The presence of four different microcyclamide-like peptides in two strains of *Microcystis* strongly suggested an abundant occurance of microcylamides or related cyclopeptides in these bloom-forming bacteria. We therefore screened 20 additional *Microcystis* strains for the presence of microcyclamide-like gene clusters (fig. 18). Primers were designed in conserved regions of the *mca*E and *mca*D genes deduced from the two known gene clusters in *M. aeruginosa* NIES298 and PCC7806. PCR analysis and subsequent sequencing for *mca*D revealed the presence of orthologues in every tested strain. However, *mca*E homologues were detectable in 14 strains.

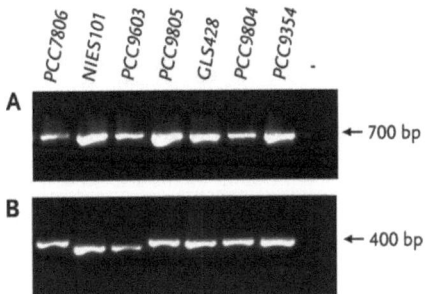

Fig. 18 PCR amplification of microcyclamide biosynthesis genes in selected *Microcystis* strains
A 700 bp fragment encoding part of the putative heterocyclisation enzyme McaD;
B ~ 400 bp fragment encoding microcyclamide precursor peptide McaE. Different length correspond to different numbers of microcyclamide repeats. Strain designations are indicated above image. (-) Water control.

The alignment of all 16 precursor peptides (fig. 19) revealed a highly conserved leader peptide and cyclisation signals, which differ only in two amino acids at two positions. Differences are detectable in size and peptide encoding regions. Actually, *Microcystis aeruginosa* NIES298 is the only strain with only two peptide encoding regions, three or four repeats seems more common. Moreover, figure 19 shows an outline of microcyclamide-like peptides encoded and reveals a high variability of amino acids at position 1, 3 and 5 of the hexapeptides, whereas position 6 is determined as cysteine.

Fig. 19 Alignment of McaE peptide sequences of different *Microcystis* strains
N-terminal leader peptides are shaded in grey; microcyclamide precursor repeats are highlighted in green. The web logo displays a scheme of the amino acid sequences of all microcyclamide encoding regions The overall height of the stack indicates the sequence conservation at that position, while the height of symbols within the stack indicates the relative frequency of each amino acid at that position.

To test the peptide repeats for diversification, we analysed the amino acid sequences using the Java application CLANS - cluster analysis of sequences (Frickey, Lupas, 2004). This method uses an algorithm to visualise pairwise sequence similarities in either two-dimensional or three-dimensional space. The CLANS analysis revealed, that the majority of peptides group in two clusters (fig. 20). The major group contains peptide repeats with the amino acid sequences HCACIC or HCATIC, respectively and comprises the sequence of the cytotoxic microcyclamide from *M. aeruginosa* NIES298. The second cluster contains variants of the sequence ATFCMC, which resemble the amino acid composition of tenuecyclamide C (ATGCMC) from *Nostoc spongiaeforme var. tenue* and nostocyclamide M (ATGCMC) from *Nostoc* 31, which both show allelopathic activity (Banker, Carmeli, 1998; Jüttner *et al.*, 2001).

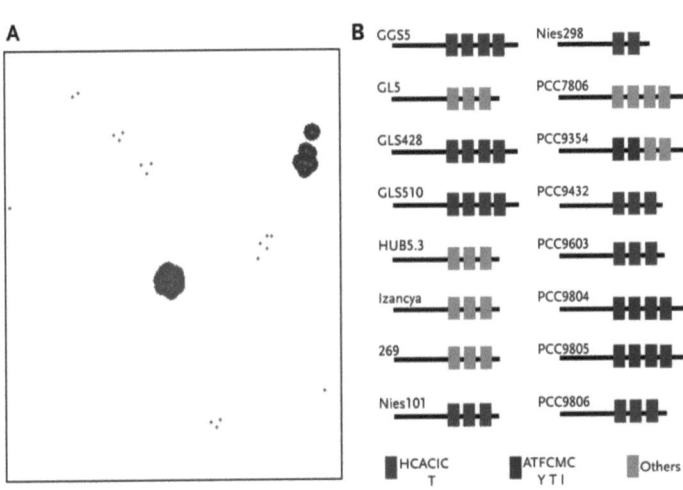

Fig. 20 CLANS analysis for peptide encoding sequences in *Microcystis* strains
A) Repeat sequences are clustered according to sequence similarity using the CLANS graphical software tool. Each node represents a particular peptide. The majority of peptide repeats assorts into two clusters, labelled in red and green, respectively. B) Schematic representation of McaE peptides showing repeats coloured according to the cluster assignment as defined in figure 3A. Sequences of repeat families including variations are shown below.

3.2 Microviridins

3.2.1 In search of the biosynthesis pathway

At the beginning of this study the common assumption was that depsipeptides are generally made by nonribosomal peptide synthetase (NRPS) assembly lines (Fischbach, Walsh, 2006; Schwarzer et al., 2003). However, extensive screening of microviridin producers for NRPS genes has not yielded candidates for the production of these tricyclic depsipeptides. Since our investigations in microcyclamide biosynthesis revealed a wide distribution of ribosomally synthesised peptides, we hypothesised a ribosomal biosynthesis pathway for microviridins as well. Our theory was supported by the fact that all known microviridins are solely composed of proteinogenic L-α-amino acids.

In parallel, Anton Liaimer from the University of Tromso performed bioinformatic analysis to search for bacteriocin-like biosynthesis genes in cyanobacteria. For that, he screened fully sequenced cyanobacterial genomes for ribosomal prepeptides (http://bioinformatics.biol.rug.nl/websoftware/bagel). He detected a putative precursor gene in *Anabaena* sp. PCC7120. The deduced gene product contains typical double glycine sequences and three copies of the KYPSD core motif that is characteristic for microviridins (fig. 21A). However, *Anabaena* sp. PCC7120 is not known to produce any microviridins and no such peptides could be detected in extracts of this strain. Hence, we decided to investigate the flanking regions of the putative prepeptide gene *all7013* to identify key features that could facilitate the search for microviridin biosynthesis genes in *Microcystis*. Sequence analysis revealed two open reading frames encoding for an ABC type transporter and an N-acetyltransferase of the GNAT family upstream of the putative microviridin precursor (fig. 21B). Moreover, we detected two open reading frames downstream of the putative precursor that, according to motif analyses (http://myhits.isb-sib.ch/cgi-bin/motif_scan), exhibit ATP-grasp ligase signatures.

The ATP-grasp family of enzymes catalyses ATP dependent formations of ester and amide bonds in peptidic structures and comprises known representatives such as D-alanine-D-alanine ligase and glutathione synthetase. We therefore considered these enzymes as excellent candidates for catalysing microviridin cyclisation reactions.

BLASTp analysis of the ATP-grasp type ligases in *Anabaena* sp. PCC7120 revealed an unnamed protein product from the marine proteobacterium *Alteromonas* sp. B-10-31 with 45% identity. This uncharacterised open reading frame is located downstream of the precursor peptide of marinostatin L (fig. 21A), a depsipeptide consisting of 11 amino acids with two internal ester linkages (Miyamoto *et al.*, 1998). Compared to microviridins, which contains three internal cross-links and the KYPSD core motif, marinostatins show, though smaller in size and lacking of an internal ω-amide bond, structural similarities. Thus, we proposed a similar biosynthetic pathway for both depsipeptides including the formation of internal ester and amide bonds by ATP-grasp type ligases.

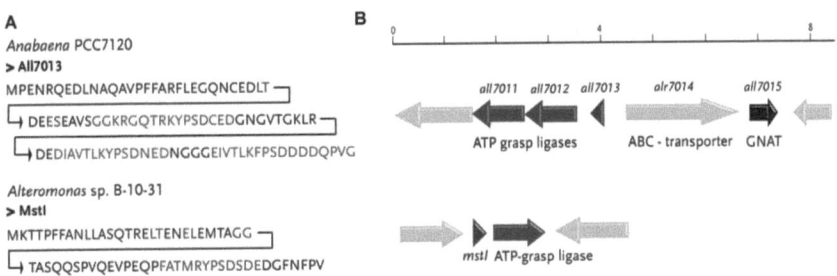

Fig. 21 Putative microviridin-like biosynthesis pathways in *Anabaena* and *Alteromonas*.
A) Deduced amino acid sequences from a putative microviridin precursor in *Anabaena* PCC7120 and from the marinostatin precursor in *Alteromonas* sp. B-10-31. Putative peptide encoding regions are shown green, deduced consensus processing sites of leader peptides and their putative double glycine cleavage sites are in red. **B)** Open reading frames in vicinity of the precursor proteins and their deduced functions. Red arrows indicate the precursor proteins, green ATP-grasp type ligases. Grey-green arrows represent ABC transporters, black ones *N*-acetyltransferases of the GNAT family.

3.2.2 Microviridin biosynthesis gene clusters in *Microcystis*

The biosynthesis of microcyclamides and patellamides showed that the processing enzymes in such a family of peptides are conserved between different cyanobacterial genera. We therefore utilised the putative ligase gene *all7011* as a probe for screening two genomic DNA fosmid libraries of the microviridin B and J producing strains *M. aeruginosa* NIES298 and MRC, respectively. Subsequent sequencing of the positive clones revealed two ATP-grasp type ligases with more than 60% identity to the *Anabaena* sp. PCC7120 enzymes in both *Microcystis* strains. Upstream of the ATP-grasp ligases, we detected open reading frames encoding putative precursor proteins. Both prepeptides are highly similar to each other except for the C terminus, which matches precisely the composition of the microviridin B (FGTTLKYPSDWEEY) or J (ISTRKYPSDWEEW) compound, respectively (fig. 22). We therefore concluded the successful identification and localisation of the microviridin biosynthesis loci in *M. aeruginosa* NIES298 and MRC and the genes were termed *mdn*.

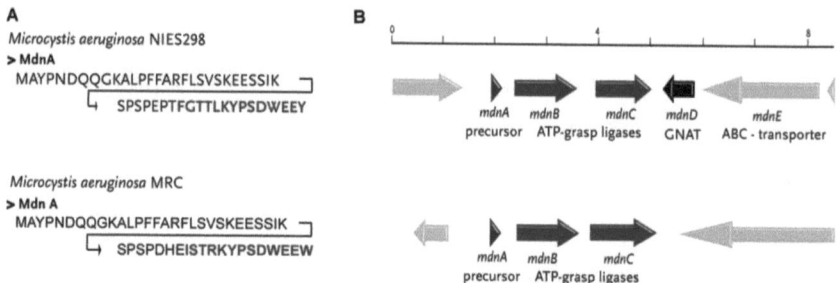

Fig. 22 Microviridin biosynthesis pathways in *Microcystis aeruginosa* NIES298 and MRC.
A) Deduced amino acid sequences from microviridin B and J precursor in *Microcystis* aeruginosa NIES298 and MRC, respectively. Peptide encoding regions are shown in green. B) *mdn* biosynthesis gene clusters of strains NIES298 and MRC. Red arrows indicate the precursor proteins, green ATP-grasp type ligases. Grey-green arrows represent ABC transporters, black ones N-acetyltransferases of the GNAT family.

The microviridin B biosynthesis gene cluster in *M. aeruginosa* NIES298 contains in addition to the precursor gene *mdnA* and the ATP-grasp ligase genes (*mdnB* and *C*) two further genes that likely code for an ABC transporter (MdnE) and an *N*-acetyltransferase of the GNAT family (MdnD) (Tab. 7). Both open reading frames are also present in the putative microviridin gene cluster in *Anabaena* sp. PCC7120. Since microviridins are commonly acetylated at the N terminus, this finding is in exact agreement with the microviridin structure. The microviridin J biosynthesis gene cluster in *M. aeruginosa* MRC neither harbours *N*- acetyltransferase nor ABC transporter genes. The two enzymes are likely encoded elsewhere in the genome, since cyanobacterial genomes are highly dynamic and generally show a very low synteny in related gene clusters (Frangeul et al., 2008).

Tab. 7 Deduced functions of ORFs in the microviridin biosynthetic gene clusters

Protein	Length (aa)	Protein-ID	Deduced Function	Sequence Similarity	Identity/ Similarity (aa length)
Microcystis aeruginosa NIES298					
MdnA	50	CAQ16116	microviridin precursor protein	hypothetical protein; *Microcystis* aeruginosa NIES843	92%/96% (50)
MdnB	325	CAQ16117	ATP-grasp type ligase	hypothetical protein; *Microcystis* aeruginosa NIES843	94%/99% (324)
MdnC	324	CAQ16118	ATP-grasp type ligase	hypothetical protein; *Microcystis* aeruginosa NIES843	98%/99% (324)
McaD	177	CAQ16119	acetyltransferase	acetyltransferase; *Microcystis* aeruginosa NIES843	97%/98% (177)
MdnE	687	CAQ16120	ABC-transporter protein	ABC-transporter; *Microcystis* aeruginosa NIES843	97%/99% (687)
Microcystis aeruginosa MRC					
MdnA	49	CAQ16121	microviridin precursor protein	hypothetical protein; *Microcystis* aeruginosa NIES843	82%/88% (50)
MdnB	327	CAQ16122	ATP-grasp type ligase	hypothetical protein; *Microcystis* aeruginosa NIES843	97%/98% (326)
MdnC	324	CAQ16123	ATP-grasp type ligase	hypothetical protein; *Microcystis* aeruginosa NIES843	98%/99% (50)

3.2.3 Heterologous expression of microviridins

To verify the involvement of the *mdn* genes in the biosynthesis and to establish a manipulable expression system, we aimed for the heterologous production of microviridins in *E. coli*. Since patellamides were directly expressed from fosmids, we first checked the *E. coli* cells harbouring the *mdn* genes in the fosmid vectors for any variations in the expression patterns. To this end, we induced the fosmids to high copy numbers and analysed the resulting cell extracts by reversed phase HPLC. Using *E. coli* cells carrying empty fosmid vectors as negative control, the metabolic profiles clearly revealed additional peaks (fig. 23). Three peaks were fractionated in extracts of the cells carrying the fosmid with the microviridin B precursor gene derived from *M. aeruginosa* NIES298, and send to Dr. Keishi Ishida at the Hans Knöll institute in Jena for further chemical analysis. MALDI-TOF analysis and NMR spectroscopy identified one peak as authentic microviridin B (fig. 23A). Two further peaks contained microviridin B variants. MALDI post-source decay analysis indicated that the peptides feature the correct ester and amide bonds. However, both compounds were not fully processed and contained one and five additional amino acids, respectively. In a similar approach, three additional peaks in the HPLC chromatogram of cells expressing the *mdn* genes with the microviridin J precursor from *M. aeruginosa* MRC were fractionated and analysed in Jena. MALDI-TOF and MALDI-PSD spectrometry demonstrated that the fractionated metabolites are derived from the microviridin J precursor and contain correct ester and amide linkages. However, none of the variants were correctly processed, all were either longer or shorter at their N termini (fig. 23B).

After isolation of the *mdn* fosmid vectors and retransformation into *E. coli* cells no expression of any peptides was detectable anymore. Hence, we considered a minimal construct, comprising only the *mdn* genes, easier to handle and manipulate (chapter 3.2.4.2)

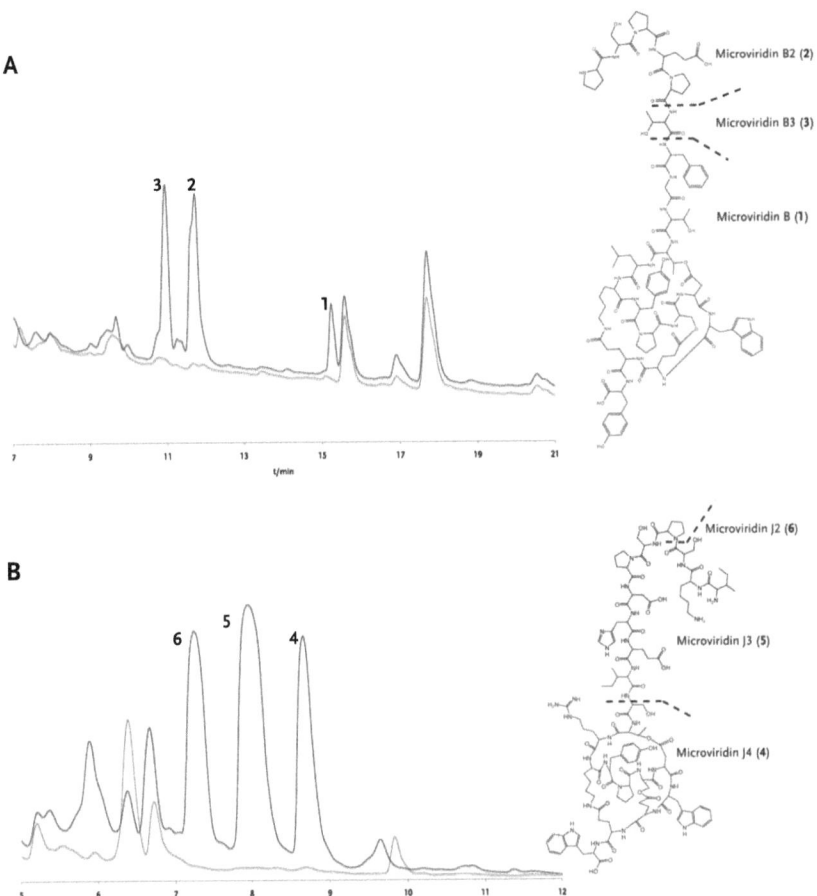

Fig. 23 HPLC profiles and structures of microviridins produced by heterologous expression
A) Heterologous expression of microviridin B pathway genes *mdnABCDE* and B) of microviridin J pathway genes *mdnABC* in *E. coli*. Extracts from expression strains are shown in red, *E. coli* with empty vectors as negative control in grey. The microviridin B and J core regions are highlighted in green.

3.2.4 Characterisation of the microviridin ligases

3.2.4.1 Phylogeny

To classify these new type of ATP-grasp ligases, we first integrated their amino acid sequences into a dataset comprising diverse enzymes of the ATP-grasp family and subjected them to a phylogenetic study. To employ two methods separately, trees were calculated with the software packages MrBayes v3.1.2. and TreePuzzle v5.2, representing the Bayesian method and a maximum likelihood approach, respectively. In both phylogenetic trees obtained, microviridin ligases fall into an independent branch, which is most closely related to glutathione S transferases (GshB) and ribosomal protein S6 modification enzymes (RimK) (fig. 24 and 25). The microviridin ligases split in two subbranches. One subbranch contains apart from All7011 from *Anabaena* sp. PCC7120 and both MdnC ligases from *M. aeruginosa* NIES298 and MRC, the enzyme that is encoded downstream of the marinostatin precursor gene in *Alteromonas*. Since marinostatins only contain ester bonds, we consider these enzymes responsible for the ester linkages in microviridin biosynthesis. The second subbranch comprising the MdnB sequences and ALL7012 from *Anabaena* sp. PCC7120 would represent microviridin amide ligases.

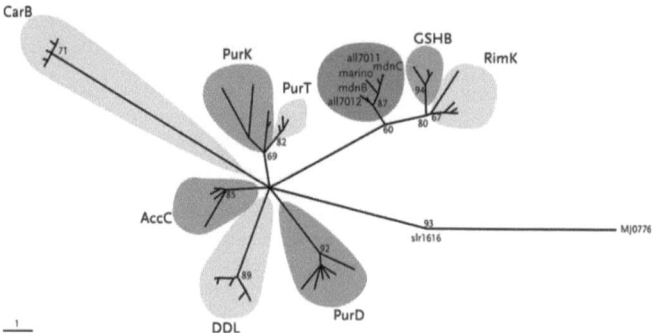

Fig. 24 Phylogenetic tree of ATP-grasp ligases using TreePuzzle.
AccC: Acetyl-CoA carboxylase; CarB: Carbamoyl-phosphate synthetase; DdlA: D-Ala-D-Ala ligase; GshB, Glutathione synthetase; PurD: phosphoribosylamineglycine ligase; PurK: phosphoribosylaminoimidazol carboxylase; PurT: phosphoribosylglycinamide formyltransferase; RimK ribosomal protein S6 modification enzyme

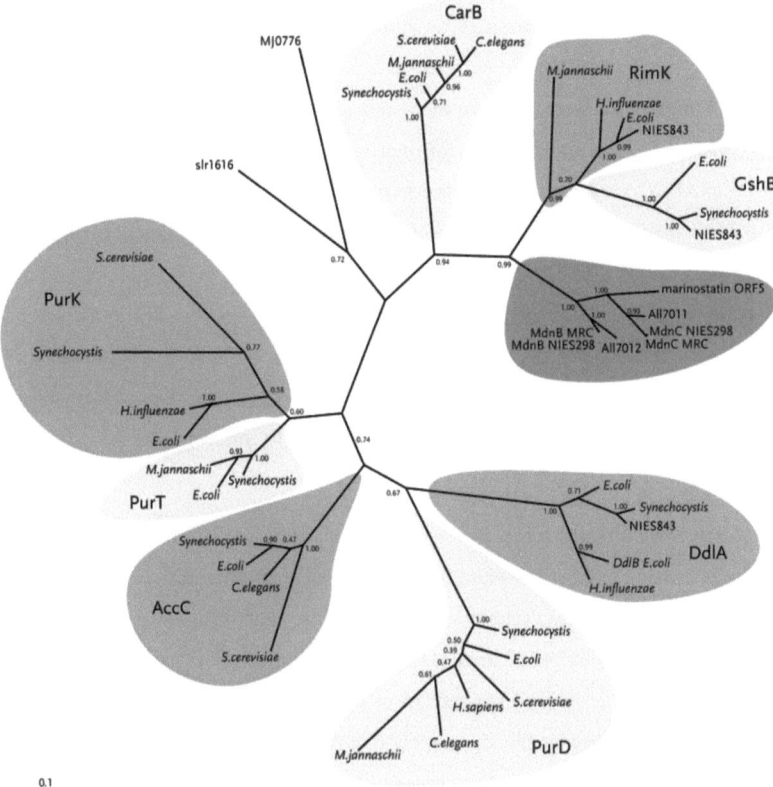

Fig. 25 Phylogenetic tree of ATP-grasp ligases as inferred by Bayesian estimation.
AccC: Acetyl-CoA carboxylase; CarB: Carbamoyl-phosphate synthetase; DdlA: D-Ala-D-Ala ligase; GshB, Glutathione synthetase; PurD: phosphoribosylamineglycine ligase; PurK: phosphoribosylaminoimidazol carboxylase; PurT: phosphoribosylglycinamide formyltransferase; RimK ribosomal protein S6 modification enzyme

3.2.4.2 Mutational analysis

In order to study the essential enzymes for microviridin biosynthesis, we aimed to perform mutational analyses. Since *Microcystis aeruginosa* NIES298 and MRC are not manipulable and first manipulation trials in *E. coli* using the fosmid expression system failed, we constructed a minimal system for microviridin production. For this purpose we cloned an *mdnABCD* gene cassette into the pDrive cloning vector (Qiagen, Hilden) and checked transformed *E. coli* cells for heterologous production of microviridin like peptides. HPLC and MALDI-PSD analysis from Dr. Keishi Ishida (HKI, Jena) revealed three correctly cyclised variants of microviridin B that only differ in the size of the *N*-terminal side chains. However, no correctly processed microviridin B was detectable (fig 26). Interestingly, microviridin producing cells generally did not survive under high selective pressure inducing higher plasmid copy numbers, possibly due to unfavourable side effects of the novel ligases.

To check whether the MdnB protein is responsible for the amide bonds in microviridins, we cleaved out a major part of the *mdn*B gene using HpaI (Fermentas, St. Leon-Rot) restriction sites and religated the vector. Cell extracts of the generated mutants lacking the *mdnB* gene were checked via HPLC analysis. No expression of any microviridin like peptides could be observed. Moreover, production of the precursor peptide MdnA in the absence of both ATP-grasp ligases failed, possibly due to rapid proteolysis of the small protein. This was confirmed by another study, where a lambda Red recombination system was used to knock out the ATP-grasp enzymes on the fosmids (Weiz, personal communication). No expression of any microviridin-like peptides was observed after retransfection in *E. coli* (see also discussion 4.4.2). Thus, we considered *in vitro* analysis an alternative approach to characterise the ATP-grasp enzymes.

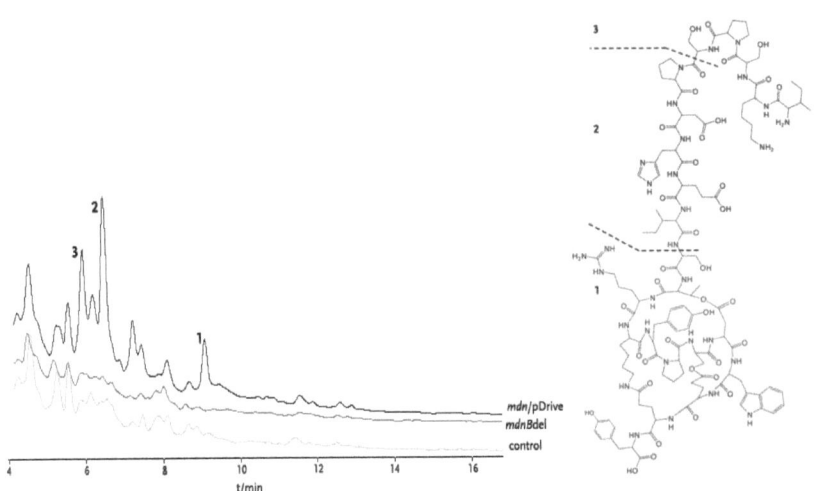

Fig. 26 HPLC profiles of extracts from *E. coli* cultures expressing *mdn* minimal constructs.
Heterologous expression of *mdnABCD* (black), *mdnACD* (grey), and negative control (light grey) in pDrive vector in *E. coli* XL-1; structures of **1-3** deduced from MALDI-TOF/PSD. The microviridin B core region is highlighted in green.

3.2.4.3 In vitro analysis of microviridin ligases

In order to perform *in vitro* assays with both ATP-grasp ligases, we designed overexpression constructs for MdnB and C in *E. coli* using the pET15b expression vector (Novagen, Nottingham). Initial irregularities in cloning and expression impeded the process. Even though we performed expression under optimised conditions, most of the enzymes were still found as insoluble aggregates. However, we finally obtained and purified soluble MdnB and MdnC (fig. 27). Overexpression of MdnA using different expression systems failed, probably due to rapid proteolysis. Therefore, we used synthetic full-length precursor peptide (MAYPNDQQGKALPFFARFLSVSKEESSIKSPSPEPTFGTTLKYPSDWEEY) and synthetic tridecapeptide precursor of microviridin B (FGTTLKYPSDWEEY) (Genscript, Piscataway) for *in vitro* assays with the putative microviridin ligases.

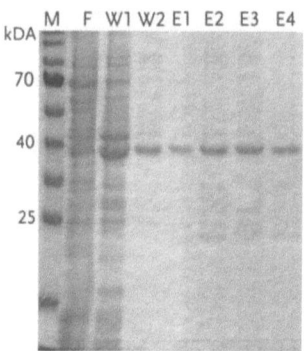

Fig. 27 **Purification of his-tagged MdnB from *E. coli* under native conditions.**
SDS PAGE with different fractions from protein purification using Ni-NTA Superflow (Qiagen, Hilden): marker (M), flow-through (F), wash fractions (W1, W2) and eluates (E1 - 4)

The enzyme assays were designed on the basis of published data of characterised ATP-grasp enzymes, such as the glutathione synthestase of *Pasteurella multocida* (Vergauwen et al., 2006) or the D-alanine-D-alanine ligase of *E. coli* (Zawadzke et al., 1991). The two overexpressed MdnB and MdnC proteins were incubated with the respective peptide substrate (10µg) in the presence of ATP. Although different buffer systems and a variety of incubation conditions were tested (see chapter 2.2.3.9), no conversion of the precursors was observed. Even cleavage of the His-tag fusion peptide did not result in any activity of the MdnB and C enzymes. HPLC chromatograms of enzyme assays with total cell lysates of the MdnB and MdnC overexpressing *E. coli* clones revealed complete degradation of the precursor, even if protease inhibitors such as PMSF were added.

In order to exclude misfolding in *E. coli*, we purified both microviridin ligases under denaturing conditions. However, subsequent *in vitro* assays with renatured and refolded enzymes failed as well. To exclude sequence mistakes in the heterologous epressed enzymes, MALDI post-source decay analysis of the purified ATP-grasp ligases were done by Keishi Ishida at the HKI in Jena, revealing that at least the primary amino acid sequence of MdnB and MdnC are correct (data not shown).

A next approach to demonstrate cylisation *in vitro*, included a second expression system. Therefore, a pACYCDuet-1 vector containing an S-tagged *mdnC* gene, was kindly provided by Annika Weiz. Expression and purification yielded sufficient pure soluble protein to perform *in vitro* assays. Several conditions were tested without any detectable transformation of the precursor peptides. In vitro assays with enriched MdnB and MdnC enzymes from extraxts of *E. coli* cells containing the fosmid with the *mdn* genes from *M. aeruginosa* NIES298 by gel filtration chromatography, failed as well.

At this time, cyclisation reactions done by MdnB and C homologues has been shown for microviridin biosynthesis in *Planktothrix agardhii* NIVA-CYA 126/8 (Philmus et al., 2008). In a similar approach, using the same protocol and a similar expression system, the authors could demonstrate that an MdnB-homologue is responsible for the amide bond and an MdnC-homologue for the ester bonds in microviridin K. Furthermore, was shown that a microviridin-precursor lacking the leader sequence is not a competent substrate for the ligases (see discussion 4.4.2).

3.2.5 Variety of microviridins

To elucidate the abundance of microviridins and check for more cryptic gene clusters, we performed tBLASTn analysis with the MdnA precursor peptides from *Microcystis* and *Anabaena*. Surprisingly, microviridins seem far more abundant than expected. In fact, we have detected related open reading frames in three more *Microcystis* strains and a number of sequenced genomes of filamentous cyanobacteria like *Nostoc* and *Nodularia* (fig. 28). Noteably, *Planktothrix agardhii* CYA126/8 contains two different precursor proteins and the unicellular cyanobacterium *Cyanothece* sp. PCC7822 even eight. However, similar ORF´s were also present in the myxobacterium *Sorangium cellulosum* SOCE56 and the marine bacteroidetes *Microscilla marina* ATCC23134 (fig. 28). The alignment of all obtained MdnA homologoues revealed two conserved regions within the proteins, comprising the microviridin core region (TxKxPSD) and one group of seven amino acids in the leader peptide (PFFARFL).

Since microviridin ligases were detectable in all of these genomes, we considered the latter region an excellent candidate for the microviridin ligase recognition site. Double glycine motifs representing putative protease processing sites are only detectable in the *Anabaena* sp. PCC7120 and in the *Microscilla marina* prepeptides.

```
                         *                ****::*                         * *;*** ;;
M.aeruginosa Nies298     MAYP--NDQQGK-------ALPFFARFLS---------VSKE-ESSIKSPSP--EPTFGTTLKYPSDWEEY-
M.aeruginosa NIES843     MAYP--NDQQGK-------ALPFFARFLS---------VSKE-ESSIKSPSP--EPTYGGTFKYPSDWEDY-
M.aeruginosa MRC         MAYP--NDQQGK-------ALPFFARFLS---------VSKE-ESSIKSPSP--DHEIS-TRKYPSDWEEW-
M.aeruginosa K139        MNYP--NSEQSK-------AIPFFARFLS---------ADQD-EAPTPDSPPDSEPAPVWTWKWPSDWED--
M.aeruginosa PCC7806     MNYP--NSEQSK-------AIPFFARFLS---------ADQD-EAPTPDSPPDSEPAPVWTWKWPSDWED--
N.spumigena CCY9414      MTTA--TLANIE-------AVPFFARFLA---------AEEPPETPAPQPEEQPLPPPIFTLKWPSDWEDC-
P.agardhii CYA126/8      MSKN-VKVSAPK-------AVPFFARFLA---------EQAV-EANNSNS-----APYGNTMKYPSDWEEY-
P.agardhii CYA126/8      MSKN-IKVSTGS-------AVPFFARFLS---------EQDT-ETGDSTSTD---IPTIWTFKWPSDWEDS-
N.punctiforme PCC73102   MPTNTVKTVDVV-------AVPFFARFLE---------EQAT-EGTEVP---------WTYKFPSDLEDR-
A.variabillis ATCC29413  MSTNTVKTVNVA-------VVPFFARFLE---------EQAT-EGTDPPS-------FPWTFKYPSDLEDQ-
Cyanothece sp.PCC7822    MSKNTGKAKEIK-------AVPFFARFLE---------EQAAQNETAPYQN--------TLKYPSDWEDY-
Cyanothece sp.PCC7822    MSKNTGKQKEIK-------AVPFFARFLE---------EQAAQNETAPYQN--------TLKYPSDWEDY-
Cyanothece sp.PCC7822    MSKNTGKAKEIK-------AVPFFARFLE---------EQAAQNETAPYV---------TKKYPSDWEEY-
Cyanothece sp.PCC7822    MSKNTGKAKEIK-------AVPFFARFLE---------EQAAQNETAPYV---------TKKYPSDWEDY-
Cyanothece sp.PCC7822    MSKNTGKAKEIK-------AVPFFARFLE---------EQAAQNETAPTV---------TRKYPSDWEDY-
Cyanothece sp.PCC7822    MSKNTGKQKEIK-------VVPFFARFLE---------EQAAQSETEALPPA-------TLKYPSDWEEY-
Cyanothece sp.PCC7822    MSKNTGKAREIRA------VVPFFARFLE---------EQAAH-PANPLP-F-------TLKYPSDWEDE-
Cyanothece sp.PCC7822    MSDINKQDASAK-------AVPFFARYLE---------EQEVSQELSQEELEG-LSGARTTLKYPSDSDEGD...
Anabaena sp.PCC7120      MPENRQEDLNAQ-------NVPFFARFLE---------GQNCEDLTDEESEAVSGGKRGQTRKYPSDCEDGN...
S.cellulosum SOCE56      MADIDNRELEQQEQNDDAEAVPFFARFLE---------DQKRVRTGVKAGRP-----PFQTLKYPSDQEDGG...
M.marina ATCC23134       MKKVK--------------KPFFAQFLENQIADEKLTNTKGGASAAAASDKKKKIKEQTMKYPSDADEDF...
                         1.......10........20........30........40........50........60........70..
```

Fig. 28 Alignment of putative microviridin precursor proteins from the database.
The microviridin core sequences are shaded in green, high conserved regions in the leader peptides are in red.

With two *mdn* biosynthesis gene clusters from *Microcystis* species at hand we designed PCR primers for the localisation of orthologous *mdn* gene clusters in a collection of these unicellular bloom-forming cyanobacteria. In fact, in all of the 18 *Microcystis* strains investigated we detected genes with high similarity to the putative microviridin amide ligase gene *mdnB*. We next analysed the same strains with a second PCR approach targeting the precursor gene *mdnA* and identified 9 candidate prepeptide genes. The amino acid sequences deduced from the sequences of the PCR products all showed the characteristic KYPSD core motif as part of a peptide coding region at their C-terminus. Six of the microviridin precursors were novel and differed in up to four amino acid positions compared to known variants.

The same approach was applied to field samples of *Microcystis* blooms from the Braakman reservoir (Netherlands) and from a bay on the Baltic Sea in Mecklenburg-Western Pomerania (Germany), kindly provided by Arthur Guljamow and Martin Hagemann, respectively. To circumvent the longsome process of *Microcystis* strain isolation and culturing, a metagenomic technique was applied. DNA was isolated directly from the filtered field samples by Arthur Guljamow and subsequently used for PCR analysis. The PCR products were cloned into the pDrive vector (Qiagen, Hilden) and transformed into *E. coli*. Subsequent sequencing of approximately 50 clones revealed seven further microviridin precursor variants (fig. 29A). At the time of sampling, *Microcystis* was the most abundant cyanobacterium in all three ecosystems. Moreover, comparison of the almost identical leader peptides (Appendix) led us to the conclusion that the obtained precursor peptides originate from *Microcystis* species.

More than half of the analysed clones from the field samples carried prepeptides encoding microviridins with the amino acid sequence YNVTLKYPSDWEEF, which is not characterised by now (fig. 29C). Approximately 10% encode for microviridin like peptides with the amino acid sequence YGVTLKYPSDWEEF and YSTLKYPSDWEEF, respectively. The only known microviridin sequence detectable in the field was the microviridin B amino acid sequence FGTTLKYPSDWEEY. To display the consensus sequence of microviridins in the field, a sequence logo was designed (fig. 29B), that clearly reveals a high variability at the first three to four positions. The core motif (TxKxPSD) was present in all detected prepeptides.

To demonstrate that screening approaches of biosynthesis genes can guide the discovery of novel peptide variants using the predicted mass of the peptides and their amino acid composition, Keishi Ishida and Christian Hertweck at the HKI in Jena isolated a new microviridin from the strain *Microcystis* NIES100 and elucidated its structure by HRFABMS and NMR spectral analysis (unpublished data). The structure indeed corresponds to the microviridin precursor sequence detected for the strain.

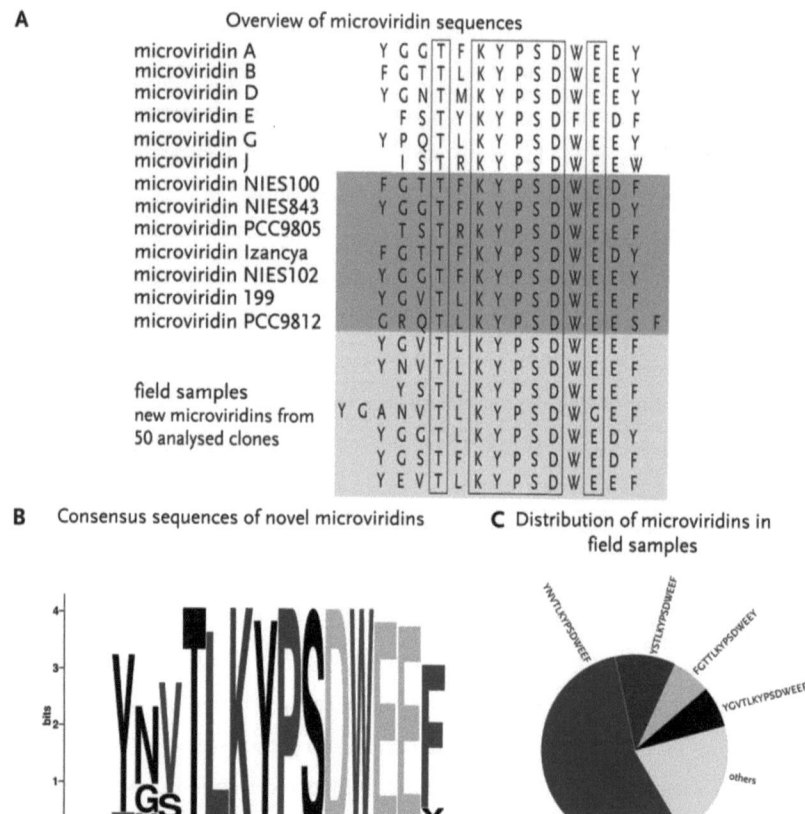

Fig. 29 Abundance of microviridin pathways in *Microcystis*.
A) Known and deduced amino acid sequences of microviridins in different *Microcystis* strains. A selection of known microviridin amino acid sequences is shown, novel microviridin sequences derived from *microcystis* lab strains are shaded in green, those from field samples in green-gray. Conserved amino acids are framed. B) The Weblogo, consists of stacks of symbols, one stack for each position in the sequence. The overall height of the stack indicates the sequence conservation at that position, while the height of symbols within the stack indicates the relative frequency of each amino acid at that position. C) The Circle diagram displays the relative abundance of deduced microviridin sequences from field samples.

3.2.6 Preliminary characterisations of the putative ABC transporter MdnE

To check for a possible role of the transporter MdnE in peptide processing, we performed a number of initial analyses. BLAST searches revealed significant similarities to putative ABC transporters present in various nonribosomal peptide biosynthesis gene clusters such as NosG from the nostopeptolide A cluster in *Nostoc* sp. GSV224 (Hoffmann *et al.*, 2003), AerN from the aeruginosin gene cluster in *Microcystis aeruginosa* NIES98 (Ishida *et al.*, 2009) or McyH from the microcystin biosynthesis gene cluster in *Microcystis aeruginosa* PCC7806 (Pearson *et al.*, 2004). None of these proteins is characterised.

We then aligned the known putative transporters encoded in microvirdin clusters from *Microcystis aeruginosa* NIES298 (MdnE), *Planktothrix agardhii* CYA126/8 (MvdA) and *Anabeana* sp. PCC7120 (Alr7014). The alignment revealed particular differences at the termini of the proteins. An additional 80 amino acid sequence is present at the C terminal region of the MdnE protein in *Microcystis aeruginosa* NIES298 (Fig. 30). BLAST analysis of this C terminal part showed its presence only in a few ABC transporters associated with cyanobacterial non-ribosomal peptides. The transporter encoded by the cryptic microviridin cluster in *Anabaena* sp. PCC7120 shows characteristic signatures of C39-type peptidases at the N terminus (Fig. 30). These domains are found in a wide range of ABC transporters, which are maturation proteases for peptide bacteriocins that cleave leader peptides at a highly conserved double glycine motif. However, whereas the putative microviridin precursor in *Anabeana* sp. PCC7120 and the marinostatin precursor in *Alteromonas* sp. B-10-31 contain typical double glycine sequences, no such markers for peptide processing are detectable in neither *Microcystis aeruginosa* nor *Planktothrix agardhii* microviridin precursors. Even so, we found that microviridin is only correctly processed *in E. coli* cells expressing the entire set of *mdnABCDE* genes, and in the absence of *mdnE* only misprocessed variants are observed. We therefore presume a so far unknown processing site for these microviridin like peptides.

N terminus

```
MdnE    ------------------------------------------------------------MPQYTTKQATENPVVSDQSANNPF
MvdA    ------------------------------------------------------------------MQTQVPPDKTITNPL
Alr7014 MKYQIVLQHSEEDCGAACIATIAKYYKRDFAIARVREAVGTGAQGTTLVGLRRGAESLGFHARQVKATPQLINQLQEAPL
        1.......10........20........30........40........50........60........70........80

MdnE    SLFGQFW----------ESLKIVAQP-----YWYPTELNG----------------------RAFGDVIISWGMSA...
MvdA    SSLTRFL----------QDVKIIAQP-----YWYPTELNG----------------------RAFSDVICSWGMLA...
Alr7014 PAIIHWKGYHWVVLYGQKGKKYVIADPGVGIRYINREELATGWANCIMLLLVPDEIRFYQQESDKIKGFSHFLARVIPYR...
        ........90.......100.......110.......120.......130.......140.......150.......160
```

C terminus

```
MdnE    ...ILDEATSALDLANEASLYQQLQESE--TTFISVGHRESLFNYHRWVLELTENSHWQLSTVEDYQRKKVNNLVMMSKKSEN
MvdA    ...ILDEATSALDLINEESLYQQLQQTQ--TTFISVGHRESLFNYHQWVLELAENSRWQFLSVEDYQQQKFGAINFLKKS---
Alr7014 ...ILDESTSALDPVLEAQVLDKLLFHRQGKTTIIISHRSRVILRADLIVYLDKGHLKLQGTLDQIQVISGEHLDFLNP----
        .......650.......660.......670.......680.......690.......700.......710.......720

MdnE    FSSVKIETLSEVKTEHDAEIAGVSEIAEGLSHQQMQSLTDYSLSNVRSRASLGKIITAKDGFNYRYDKDPKTLKWVRI
MvdA    -----------------------------------------------------------------------------
Alr7014 -----------------------------------------------------------------------------
        .......730.......740.......750.......760.......770.......780.......790.
```

Fig. 30 Alignment of the ABC type transporters from different *mdn* gene clusters.
MdnE from *Microcystis aeruginosa* NIES298, MvdA from *Planktothrix agardhii* CYA126/8 and Alr7014 from *Anabaena* sp. PCC7120. The predicted C39-type peptidase domain is shown in green, specific N terminal sequences are shaded in grey-green.

Since these data strongly suggest that the transporter MdnE is essential for a correct processing of microviridin like peptides, we tried to perform a *trans* complementation of the missing *mdnE* in the microviridin J biosynthesis gene cluster. Therefore, we cloned the *mdnE* transporter gene including the putative promotor region into the pACYC184 plasmid vector (MBI Fermentas, St. Leon-Rot) and transformed it into the respective *mdn* fosmid containing *E. coli* cells. Unfortunately no change in peptide expression was detectable. We then decided that further characterisation of MdnE, no matter how interesting and important, would go beyond the scope of this study and deserved closer attention in a project of its own. Weiz and colleagues (personal communications) could show that coexpression of the microviridin B containing fosmid with the MdnE transporter in the pET expression vector, clearly increases the yield of authentic microviridin B. Furthermore, after coexpression of the transporter and the *N*-acetyltransferase in the pET system with the fosmid containing the microviridin J biosynthesis genes, authentic microviridin J was detectable (see discussion 4.4.3).

3.2.7 Microviridin expression studies in cyanobacteria

3.2.7.1 Generation of an antibody against MdnB

To perform expression studies in *Microcystis* and *Anabaena* an antibody against one of the ATP-grasp ligases was generated. For that, his-tagged MdnB protein from *M. aeruginosa* NIES298 was expressed in *E. coli* and purified under denaturing conditions. The purified protein was used to raise a polyclonal rabbit antibody (Pineda-Antibody-Service, Berlin). To test the antibody, immunoblots with expressed proteins from *E. coli* and cyanobacterial protein extracts were performed every 30 days. After 120 days of incubation the antiserum showed a highly specific reaction against the MdnB protein, in the overexpressing *E. coli* strain (fig. 31). Specific signals could also be detected in protein extracts of *M. aeruginosa* NIES298 and MRC and *Anabaena* sp. PCC7120. However, the MdnB protein detectable in *M. aeruginosa* MRC has a size of approximately 33kDa and seems therefore smaller than the expected 37 kDa (Fig. 32A, chapter 3.2.7.2). In spite of high similarities between the two microviridin ligases, western blot analysis with heterologous expressed and purified MdnC protein revealed only weak signals (data not shown).

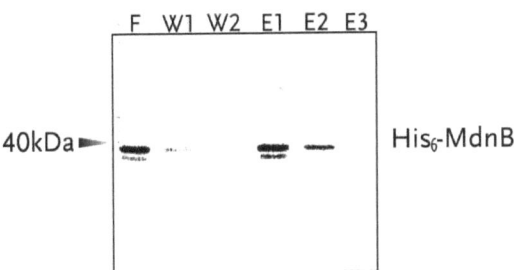

Fig. 31 Immunoblot analysis of Purification of his-tagged MdnB using an anti-MdnB antibody.
Western Blot with different fractions from protein purification of his-tagged MdnB under denaturing conditions: marker (M), flow-through (F), wash fractions (W1, W2) and eluates (E1 - 3). The additional smaller band was supposed as specific degradation product of the his-tagged ligase.

3.2.7.2 Immunodetection of MdnB in Microcystis and Anabaena species

With the anti-MdnB antibody available, we tested the abundance of the protein in different cyanobacterial cell extracts at various conditions. First experiments revealed major differences in MdnB protein levels in all tested strains with no relation to particular light intensity (data not shown) or cell densities (fig. 32). Fractionating the protein extracts into periplasma, membrane associated proteins and cytosol, showed that most of the MdnB protein was detectable in the cytosolic fraction (fig. 32B). Variations in light intensities and cell densities resulted in an incidential expression pattern. Possibly, further specifications are needed to get a stable MdnB expression. Northern Blot analysis of the *mdnA* precursor showed similar incidental transcription patterns (data not shown). However, if the protein was detectable, it was of high abundance. In the *Anabaena* strain PCC7120 more than the expected 37kDa band was detectable (fig. 32C). Since the antibody is very specific, at least the lower band seems to be a degradation product. To test a time-dependent expression of the microviridin ligase, we incubated cell cultures of *Anabaena* sp. PCC7120, *M. aeruginosa* NIES298 and MRC at day and night cycles of 16 h and 8 h light, respectively. After two month of adaption and synchronisation, samples were taken every three hours for one day, once in the exponential and in the stationary phase. In all samples no MdnB protein was detectable (see discussion).

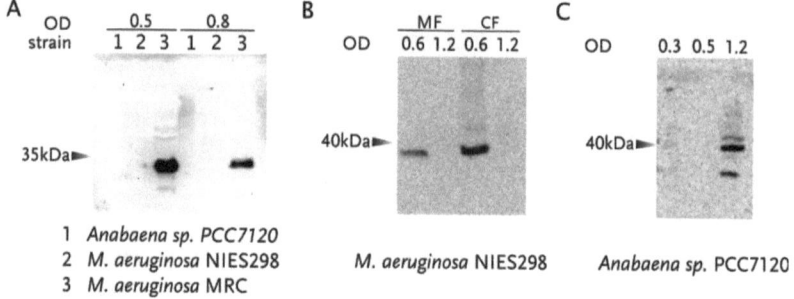

Fig. 32 Immunoblot analysis of cyanobacterial cell extracts using an anti-MdnB antibody.
A) Expression of MdnB in cell extracts of three microviridin encoding cyanobacterial strains: *Anabaena* sp. PCC7120, *M. aeruginosa* NIES298 and MRC, at two different cell densities (OD$_{750}$). B) Expression of MdnB in *M. aeruginosa* NIES298 at different cell densities (OD$_{750}$) in membrane (MF) and cytosolic (CF) fractions. C) Expression of MdnB in cell extracts of *Anabaena* sp. PCC7120 at different cell densities (OD$_{750}$).

3.2.7.3 Visualisation of MdnB by immunofluorescence microscopy

Immunofluorescence is a common method to visualise structures and localise proteins in bacterial cells. With the specific anti-MdnB antibody and a FITC-labelled anti-rabbit antibody at hand, we performed fluorescence *in situ* analysis with *Microcystis* and *Anabaena* cells from cultures grown under different light intensities at different growth phases. Controls were performed using only the secondary antibody or without any antibody showing only the green autofluorescence of the cyanobacteria. The red chlorophyll fluorescence of the cyanobacterial cells served also as a control verifying the number and localisation of viable cells under the microscope. Again, specific MdnB signals occur only incidentally. However, if signals were detectable, cells treated with the anti-MdnB antibody showed a brighter green fluorescence covering the whole cytosol of *Microcystis aeruginosa* NIES298 cells compared to the controls (fig. 33). Interestingly, we detected several times specific signals around the septum in dividing cells of *M. aeruginosa* NIES298 (fig. 33A). No such structures could be observed in cells of *M. aeruginosa* MRC or *Anabaena* sp. PCC7120.

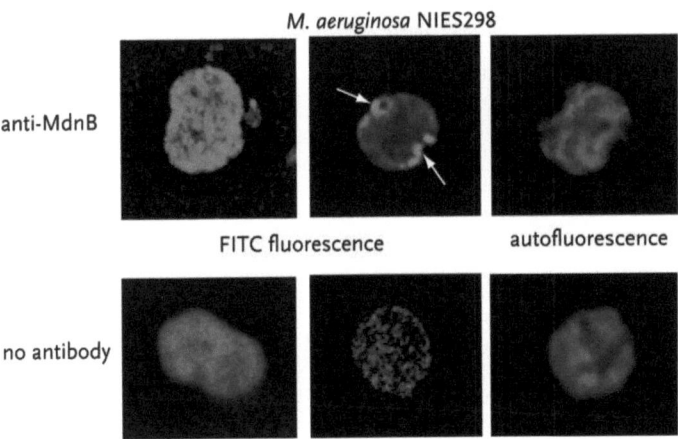

Fig. 33 **Immunofluorescence micrographs of cyanobacterial cells using the anti-MdnB antibody.** FITC fluorescence and autofluorescence of *M. aeruginosa* NIES298 cells treated with the anti-MdnB antibody and a FITC labelled secondary antibody or just the secondary antibody as control. White arrows indicate putative specific signals near the septum of a dividing cell.

4 Discussion

The bloomforming freshwater cyanobacterium *Microcystis aeruginosa* is known to produce a variety of compounds covering a broad spectrum of bioactivities. In the course of this study biosynthesis pathways of two classes of therapeutically useful peptides were identified. The cytotoxic hexapeptide microcyclamide is formed by a ribosomal pathway through the activity of a set of enzymes closely resembling those recently shown to be involved in patellamide biosynthesis. A unique ribosomal biosynthesic pathway involving uncharted ATP-grasp enzymes has been elucidated for the biosynthesis of the tricyclic depsipeptides microviridins. Their successful heterologous expression in *E. coli* provides a promising base for engineering novel variants.

Screening genome databases, *Microcystis*-laboratory strains and field samples revealed a wide spread occurrence of microcyclamide and microviridin biosynthesis genes in cyanobacteria and demonstrated a high natural diversity, that guided the discovery and structural elucidation of novel peptide variants.

Raising the question for the ecological role of these peptide families, transcription and expression studies of the biosynthestic genes provided first insights into regulation mechanisms.

4.1 Microcyclamide biosynthesis in *Microcystis*

4.1.1 A patellamide-like biosynthesis of microcyclamides

Identification of the microcyclamide biosynthesis genes in *M. aeruginosa* NIES298 revealed a ribosomal assembly line similar to those from the patellamides. The gene cluster comprises 13 kb and contains nine genes, seven of which are orthologues to patellamide genes.

Unfortunately, heterologous expression of microcyclamide in *E. coli* failed. Although shown for patellamides (Schmidt et al., 2005), no peptides could be detected in cell pellets or supernatants of *E. coli* cells harbouring the *mca* fosmid. Since *M. aeruginosa* NIES298 is not genetically manipulable, possibly due to restriction endonucleases (Frangeul et al., 2008; Takahashi et al., 1996), expression in *E. coli* would be an important base for mutational studies of the biosynthesis pathway. Maybe the absence of an associated methyltransferase in the used *E. coli* strain or the presence of the double glycine motif in the leader peptide hindered the production of microcyclamide. The second hypothesis is supported by observations made during the expression of microviridins in *E. coli*. Heterologous expression from the fosmids worked only for microviridin B and J from *Microcystis* strains, the corresponding *Anabaena* microviridin containing a double glycine motif in the precursor peptide was not expressed in *E. coli*. Many different reasons could be responsible for the difficulties in heterologous expression of peptides formed from precursors with a double glycine motif. Differences in cell wall composition could be one reason for the difficulties resulting in a loss of function of putative transporter protease domains. ABC transporters associated to processing and export of peptide pheromones such as colicin from *E. coli* are usually linked to membrane fusion proteins and outer membrane proteins such as TolC (Michiels et al., 2001). Since the peptidoglycan layer of cyanobacteria is considerably thicker (Hoiczyk, Hansel, 2000), interactions of cyanobacterial ABC transporters and transporter-associated proteins from *E. coli* might be impossible. However, further investigations are needed to clarify these questions.

Due to the similarity to the biosynthesis enzymes of patellamides, an analogous pathway to form microcyclamide seems most likely. As Schmidt and colleagues already characterised enzymes of the patellamide pathway, no further efforts in that direction regarding microcyclamides were made. Nevertheless, sequence comparisons of precursor and modifying enzymes can reveal interesting insights to biosynthetic mechanisms.

The precursor protein McaE forms the substrate for posttranslational modifications by McaA, McaD, McaF and McaG (fig.34). McaA, analogous to PatA, catalyses the proteolytic cleavage of the N-terminal recognition sequence G(L/A)EAS in the McaE precursor peptide (Lee et al., 2009). The product of McaA is used as substrate for McaG. As shown for PatG (Lee et al., 2009), McaG is predicted to catalyse the cleavage of the C-terminal sequence AFD in tandem with cyclisation of the microcyclamide peptide. A one-step transamidation mechanism was proposed to explain the cyclisation reaction (Lee et al., 2009).

Although not characterised, according to the similarity to the SagC and D families of enzymes, it is suggested that McaD is responsible for the heterocyclisation of cysteine and serine or threonine into thiazoline and oxazoline rings. The SagBCD proteins and their homologues have been shown to introduce heterocycles onto precursor peptides of a number of bacteriocins such as microcin B17 and streptolysin S (Lee et al., 2008). Especially the microcin B17 biosynthesis has been subject of extensive studies and contributes to the understanding of the basic principles of heterocyclisation in small ribosomal peptides (Li et al., 1996; Milne et al., 1998; Sinha Roy et al., 1999). Accordingly, McbB a SagC-homologue and zinc binding protein, is responsible for the heterocyclisation reaction. The putative docking protein McbD, a SagD homologue including an ATPase/GTPase domain, is also required for compound production, although the exact mechanism is still elusive. In microcin B17 biosynthesis, a third protein is reqired for the heteroclisation reactions, McbC. This microcin B17 oxidase is a flavoprotein and is thought to oxidise thiazolines and oxazolines to thiazoles and oxazoles. Since no such domain is detectable in the McaD protein, the mcbC-like oxidoreductase in the N terminus of the McaG protein is predicted to play that role in the microcyclamide biosynthesis.

Another enzyme is suggested to participate in heterocyclisation as well. The PatF protein was shown to be essential in patellamide biosynthesis, but is absent in the trichamide gene cluster. As no oxazoline is part of the trichamide peptide, PatF was proposed to be involved in oxazoline formation (Sudek et al., 2006).

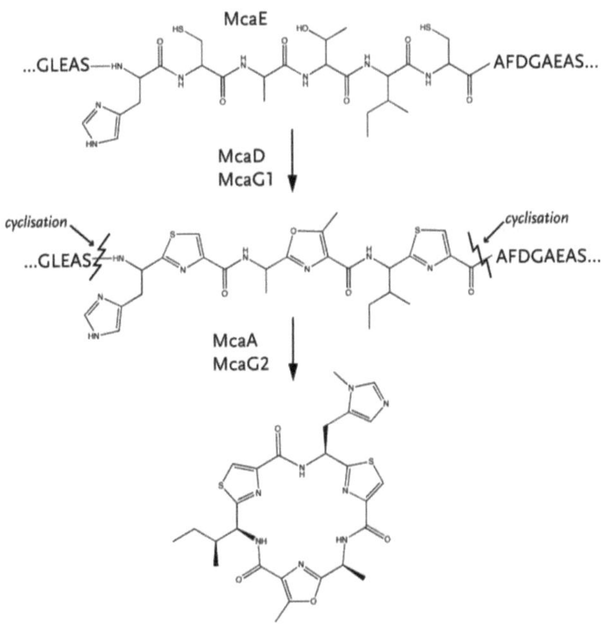

Fig. 34 Proposed pathway to microcyclamide in *M. aeruginosa* NIES298.

No functional role could be assigned to McaB and McaC. Both proteins have been shown to be dispensable in patellamide biosynthesis. However, homologues are present in all known patellamide-like gene clusters, suggesting a specific role in the biosynthesis of these small cyclic peptides. Furthermore, the role of the two additional ORFs remains unclear. Both are not present in any other patellamide-like gene cluster known so far. Due to that fact and their small size, we consider them as pseudogenes. The microcyclamide gene cluster does not code for a protein possessing methyltransferase domain signatures that could be responsible for the methylation of histidine in microcyclamide. Therefore, it remains ambiguous whether one of the uncharacterised enzymes in the gene cluster or factors encoded in *trans* provide the required methylation activity. The epimerisation mechanism is still under discussion as well. It was proposed that epimerisation of single amino acids within patellamides occurs spontaneously and is interdependent with macrocyclisation of the linear prepeptide. This may also explain the fact that different patellamide and cyclic hexapeptide variants were shown to differ in the stereochemistry of individual amino acids (Banker, Carmeli, 1998; Degnan *et al.*, 1989; Ishida *et al.*, 2000; Jüttner *et al.*, 2001).

Taken together, a detailed analysis of the microcyclamide gene cluster in *M. aeruginosa* NIES298 has revealed the expected similarity but also clear differences from the recently described patellamide and trichamide gene clusters. This study provides the first evidence for the biosynthesis of a cyclic hexapeptide from a ribosomal precursor in cyanobacteria and suggests a similar pathway for all cyclamides.

4.1.2 Shedding light on the precursor

Interesting similarities emerge when comparing of the precursors of known patellamide like peptides. Although size and number of the peptide coding regions differ, the protease recognition sequences are conserved. Taking into account the high similarity of the protease and cyclisation enzymes, similar recognition sequences make sense. The leader peptides contain regions with highly conserved amino acids. The first five amino acids are identical in all precursor peptides and a nine amino acid domain (LAELSEEAL) at the end of the leader peptide is detectable in four of the five precursors analysed. One or both motifs could serve as recognition sequence for the docking domain of the McaD protein to guide heterocyclisation. The presence of a double glycine at the end of the leader peptide of the microcyclamide precursor suggests a possible transport through the inner cell membrane, similar to bacteriocin biosynthesis. Conserved motifs in the leader peptide could then serve as signal sequences for the peptidase domain of the transporter.

Against that hypothesis argues the fact that no other precursors contain the double glycine sequence. Furthermore, consensus sequences in front of the double glycine, deduced from different bacteriocins in gram-positive and gram-negative bacteria LSX₂ELX₂IX(GG) (Michiels *et al.*, 2001), show no similarity to the LAELSEEAL(GG) motif in front of the microcyclamide double glycine. No microcyclamides were detected in the supernatant of *Microcystis aeruginosa* NIES298 cultures and no typical ABC transporter is present in the gene cluster, appearance of the GG in the leader peptide may be incidental. Nevertheless, patellamides were detectable in the ascidian hosts of their bacterial producers, suggesting a possible transport through the cell membrane.

PatE (*Prochloron* sp.)
MNKKNILPQQGQPVIRLTAGQLSSQLAELSEEALGDA<u>GLEAS</u>VTACITF<u>ADGVEPS</u>JTVCISVC<u>AYDGE</u>

TenE (*Nostoc spongiaeforme* var. *tenue*)
MDKKNILPQQGKPVIRITTGQLPSFLAELSEEALGDA<u>GVGAS</u>ATGCMC<u>AYDGAGAS</u>ATGCMC<u>AYDGAGAS</u>ATACAC<u>AYDGAGAS</u>ATACAC<u>AYE</u>

McaE (*Microcystis aeruginosa* NIES298)
MDKKNILPQQGKPVFRTTTGKLPSYLAELSEEALGGN<u>GLEAS</u>HCATIC<u>AFDGAEAS</u>HCATIC<u>AFDGDEA</u>

McaE (*Microcystis aeruginosa* PCC7806)
MDKKNLLPNQGAPVIRGISGKLPSHLAELSEEALGGN<u>GAEAS</u>ATVSIC<u>AFDGAEAS</u>FTGCMC<u>AFDGAEAS</u>ITGCIC<u>AFDGAEAS</u>ITGCIC<u>AFDGDEA</u>

TriG (*Trichodesmium erythraeum*)
MGKKNIQPNSSQPVFRSLVARPALEELREENLTEGNQGHGPLAN<u>GPGPS</u>GDGLHPRLCSC<u>SYDGDDE</u>

Fig. 35 Selected cyanobactin precursor peptides.
Peptide encoding regions are shown in green, protease recognition sequences are underlined, putative double glycine motifs are shown in red.

4.1.3 Diversity of microcyclamides

At the beginning of this study only two similar gene clusters were known, one responsible for the patellamide biosynthesis in *Prochloron* species (Schmidt *et al.*, 2005) and the other for the trichamide biosynthesis in the marine free-living cyanobacterium *Trichodesmium erythraeum* (Sudek *et al.*, 2006). Apart from the microcyclamide genes, further similar gene clusters in cyanobacteria were discovered and published within the time of this study.

Similar biosynthetic pathways could be assigned to tenuecyclamides, trunkamide and patellins (Donia et al., 2008), demonstrating the natural diversity of these assembly lines. Since all these pathways have conserved features, they were included in a new family of cyanobacteria-specific compounds called cyanobactins (Donia et al., 2008).

The cyanobactins are low-molecular weight cyclic peptides, which can contain heterocyclised or prenylated amino acids. About 100 probable cyanobactins have been previously described and more await discovery. In this study the wide spread occurrence of cyanobactins within the genus *Microcystis* has been shown. The identification of a second gene cluster in the genome of *M. aeruginosa* PCC7806 revealed highly conserved modification enzymes, only the peptide coding regions in the precursor differed in number and amino acid sequence. Three different peptides were encoded by four repeats of the peptide coding cassette and all three microcyclamide variants were shown to be expressed (Portmann et al., 2008a; Portmann et al., 2008b). Screening of 20 additional strains, isolated from all over the world (tab. 2, chapter 2.1.1), revealed a far larger genetic potential of cyclamide production in *Microcystis* than expected, as only one microcyclamide from this genus was characterised so far. In every strain tested at least a McaD homologue could be detected, suggesting similar gene clusters to be present in every strain.

A recent screening of 132 taxonomically and morphologically diverse free-living cyanobacterial strains for *mca*A orthologues, identified genes in 48 strains of unicellular, filamentous and heterocyst differentiating cyanobacteria and underlined the wide distribution of these cyclic peptides (Leikoski et al., 2009). Cyanobactin biosynthesis genes were detected in a number of strains including the genera *Microcystis*, *Anabaena*, and *Nostoc* and appeared to be especially common in planktonic, bloom forming cyanobacteria. Furthermore comparison of phylogenetic trees of 16S rDNA genes and cyanobactin genes revealed that horizontal gene transfer (HGT) events were likely to play a role in the dispersal of these metabolites among cyanobacteria. HGT is an important mechanism for the evolution of microbial genomes. Especially secondary metabolites are prime candidates for HGT, because they are not essential to the microbial cell and may provide new physiological functions (Dobrindt et al., 2004).

Various biosynthesis gene clusters of secondary metabolites have been shown to lie within so called genomic islands, mobile genetic elements, which have been shown to be involved in HGT. Gene transfer seems not only possible between microbes, but also between eukaryotes and bacteria. A recent study revealed, that the eukaryotic cytoskeletal elements actin and profilin were transfered into the cyanobacterium *Microcystis* (Guljamow *et al.*, 2007). According to the higher GC content compared to the genetic background, the patellamide biosynthesis genes were suggested to be transferred into *Prochloron* via HGT (Schmidt *et al.*, 2005).

Number and size of the peptide coding regions in different cyanobactin precursors differ. Whereas the trichamide precursor peptide encode for only one 11-amino acid peptide (Sudek *et al.*, 2006), patellamide and lissoclinamide precursors in *Prochloron* strains have been shown to encode two peptides of seven or eight amino acids (Donia *et al.*, 2006). Up to four peptide coding regions are detectable in McaA homologues encoded by *Microcystis* strains, all encoding for a six amino acid peptide. The diversification of the patellamide pathway in *Prochloron* and the presence of various patellamide-like peptides in the host ascidians led to speculations about an immune-like function (Donia *et al.*, 2006). The evolutionary pathway of the patellamides, by shuffling hypervariable peptide coding cassettes in a conserved genetic environment, was compared to the acquired immune system of eukaryotes. The model of patellamide diversification involved an ancestral duplication event of a peptide encoding cassette including the surrounding signal sequences. Subsequently these sequences rapidly diversified either by directed mutations or by recombination events. A similar scenario can be assumed for the microcyclamide diversification.

Since the posttranslational machinery has been shown to be highly tolerant of diverse substrate sequences (Donia *et al.*, 2006; Lee *et al.*, 2009), the evolution of different cyanobactins only requires a switch in the small peptide coding cassettes. In addition to NRPS and PKS (Jenke-Kodama, Dittmann, 2009a; Jenke-Kodama, Dittmann, 2009b; Schwarzer *et al.*, 2003), the cyanobactins therefore represent a new example of diversification processes of secondary metabolites.

4.1.4 Towards the role of microcyclamides

Although no functional role could unambiguously be assigned to microcyclamides or cyanobactins in general, comparison of precursors and related pathways could help to understand their variety and function. The high diversity of these peptides in cyanobacteria suggest a possible role in communication or chemical defence mechanisms.

The cluster analysis of the peptide coding regions clearly demonstrate two major groups of microcyclamide variants in *Microcystis*. The biggest group contains the amino acid sequence HCA(C/T)IC (Ishida et al., 2000), and therefore comprises the cytotoxic microcyclamide from *M. aeruginosa* NIES298. The second group contains the consensus sequence AT(F/Y)(C/T)(M/I)C and shows some sequence similarities to the allelopathic compounds tenuecyclamide C and nostocyclamide M (ATGCMC) from different *Nostoc* species (Jüttner et al., 2001). An allelopathic function of these peptides in *Microcystis* can therefore be speculated. A common theory about the production of secondary metabolites is their function as agents of "chemical warfare" (Baba, Schneewind, 1998) (Demain, Fang, 2000). Known plant derived chemical defence metabolites are for example nicotine (Winz, Baldwin, 2001) and pyrrolizidine alkaloids (Hartmann, 2004).

The production of toxic compounds against biological competitors in bacteria has been discussed especially for the known antibiotic producing *Streptomyces* species (Challis, Hopwood, 2003; Maplestone et al., 1992) but was also described for cyanobacteria. Lyngbyatoxin A and cyanobacterin are two examples for cyanobacterial compounds with allelopathic activity (Paul et al., 2007) (Berry et al., 2008). The fact that not all cyanobactins show any inhibitory activity, argues against a general antibiotic function.

According to the "Screening Hypothesis" of Firn and Jones, organisms have to generate considerable chemical diversity, to enhance their chances of finding a novel beneficial chemical, as potent biomolecular activity is a rare property for any compound to possess (Firn, Jones, 2000; Firn, Jones, 2003; Firn, Jones, 2006). A central postulate of the Screening Hypothesis is that it is the overall capacity to produce secondary metabolites that is shaped by evolution. In simple terms, organisms produce a high chemical variety in order to find one useful. The possession of the overall machinery is crucial, but most substances made by that machinery confer no advantage to the producer. Although that theory is highly disputed, it would explain the differences in antibiotic and cytotoxic activities of canobactins, suggesting that non-antibiotic acting variants are incidentally produced to maintain chemical diversity.

Other theories deny a general antibiotic function of these metabolites and suggest an important role in cell-cell communication (Davies, 2006). Their main argument is based on the fact that concentrations needed to show antibiotic effects are rarely achieved. Furthermore, has been shown, that known antibiotics such as rifampicin and erythromycin in subinhibitory concentrations can alter global transcription patterns in bacteria (Davies, 2006). Mutagenesis of biosynthetis genes of yellow pigments called DK xanthenes in *Myxococcus xanthus* delayed fruiting body formation and sporulation (Meiser et al., 2006). A possible role in colony formation was proposed for the toxin microcystin in *Microcystis* (Kehr et al., 2006). A lectin involved in cell-cell recognition and cell-cell attachment has been shown to be differentially expressed in microcystin deficient mutants. Several of the peptide bacteriocins of gram-positive bacteria like subtilin, nisin, and other lantibiotics are primarily produced in the late growth phase and act as peptide pheromones, wich induce their own biosynthetic genes and other target genes (Kleerebezem, 2004). Therefore, we suggested that microcyclamides could potentially be involved in communication and self-recognition of *Microcystis* ecotypes. As transcription analysis of *mcaE* showed similar transcript levels independent from cell growth phase or light intensity, it seems unlikely that microcyclamides act as autoinducers of their biosynthesis genes, as this should have been reflected by cell density-dependent expression.

The alignment and consensus sequence of the identified precursor peptides from *Microcystis* (fig.19, chapter 3.1.6) revealed differences in conservation of the specific amino acid positions. Position 6 is determined as cysteine and needed as backbone amino acid for the thiazole ring formation. Position two and four can contain cysteine, threonine or serine residues, needed for the conversion into thiazole or oxazoline/oxazole rings.

Notably, several precursor sequences encode an arginine at the second position, which is not known to play a role in heterocyclisation, indicating that heterocycles at this position are not essential. That is in agreement with the fact that new microcyclamides PCC7806A and B and aerucyclamides from *M. aeruginosa* PCC7806 contain either oxazoline rings or open threonine residues at this position. However, most known cyclamides, such as nostocyclamide, tenuecyclamide, venturamide and microcyclamide contain only fully oxidised heterocycles (Ishida *et al.*, 2000; Jüttner *et al.*, 2001; Linington *et al.*, 2007). Therefore, it can be speculated that the oxidation status of the heterocycles in cyclamides is associated with their pharmaceutical activities, as cyclamides from *M. aeruginosa* PCC7806 show less potent bioactivities than other cyclamides. These structure-activity relationships have been shown for the lissoclinamides (Hawkins *et al.*, 1990). Oxidation of oxazoline and thiazoline to oxazole and thiazole respectively, significantly affected bioactivity. It remains elusive, whether this is connected with the ecological role of these peptides.

According to their high abundance in the host ascidians, patellamides were proposed to minimise predation or pathogenesis, to be part of an interstrain competition of *Prochloron* spp. within ascidians, somehow play a role in adaption of the host to different environmental conditions or act as communication molecules in host symbiont interactions (Schmidt *et al.*, 2005, Donia, 2006 #77). Although no significant biological activities were found for trichamide, an antipredator function was proposed for the peptide. Over the years, more biosynthetic clusters will be discovered and characterised and mutational studies in genetic manipulable systems could shed light on the physiological role of these peptides.

4.2 Microviridin Biosynthesis – a novel ribosomal pathway

In the course of this study two gene clusters that code for the tricyclic depsipeptides microviridin B and J were identified, cloned and sequenced. Heterologous expression of these peptides in *E. coli* confirmed that microviridins derive from a ribosomal peptide precursor, which is posttranslationally modified by tailoring enzymes including two ATP-grasp-fold ligases, an N-acetyltransferase and an ABC type transporter. The proposed biosynthetic mechanism represents a previously uncharacterised strategy for the assembly of small cyclic peptides (fig.36). Up to this point ATP-grasp ligases were not known to be involved in the biosynthesis of small cyclic peptides. Heterologous expression of a construct containing the precursor and the two ligases demonstrated that these three genes are sufficient for the production of a correctly cyclised microviridin peptide prior to cleavage of the N-terminal leader sequence. However, *in vitro* characterisation of the ATP-grasp ligases failed.

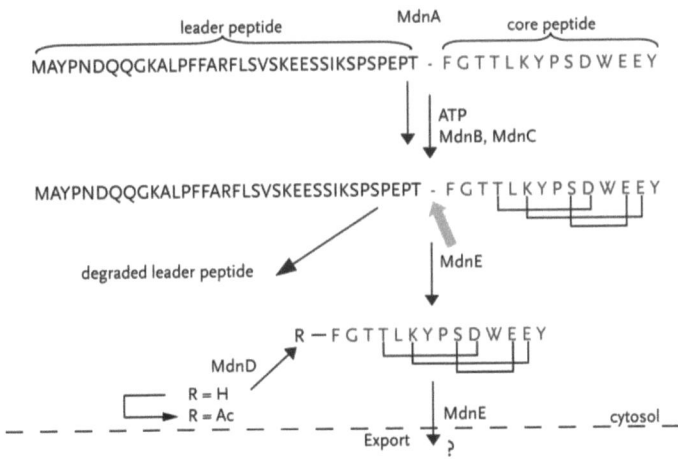

Fig. 36 Model for microviridin biosynthesis.
Cyclisation of the prepeptide MdnA by grasp ligases MdnB and MdnC prior to cleavage by transporter peptidase MdnE and N-acetylation by MdnD.

4.2.1 The microviridin ligases

The two proteins encoded adjacent to the precursor peptide belong to a diverse superfamily of enzymes that is known in both prokaryotes and eukaryotes and catalyses ATP dependent ligations of a carboxylate containing molecule to an amino or thiol group containing molecule. In the process, ATP binds in a cleft formed by two sub-domains, each containing two antiparallel β-strands and a loop. Currently, 17 sub-groups of enzymes belong to the ATP-grasp family, including known representatives such as D-alanine-D-alanine ligase, glutathione synthetase, carbamoylphosphate synthetase and the ribosomal S6 modification enzyme (RimK). The obtained phylogenetic tree (fig.25, chapter 3.2.4) reveals that microviridin ligases form their own subtree and therefore represent a new group of ATP-grasp fold enzymes, closely related to glutathione S transferases and ribosomal protein S6 modification enzymes (RimK).

Whereas ATP-grasp ligases are well characterised in primary metabolism, only a couple of similar enzymes are known for their functions in secondary metabolism. The cyanophycin synthetase, which contains an ATP binding region reminiscent to ATP grasp ligases, has been shown to catalyse the condensation of multi-L-arginyl-poly (L-aspartatic acid), so called cyanophycin - a protein-like cell inclusion, which acts as a temporary nitrogen storage in cyanobacteria (Berg *et al.*, 2000; Ziegler *et al.*, 1998). Another example is the formation of *N*-glycyl-clavaminic acid by an ATP-grasp type ligase as an intermediate in the biosynthetic pathway of the β-lactamase inhibitor clavulanic acid (Arulanantham *et al.*, 2006).

Heterologous expression of microviridin J-like variants from the fosmid derived from *M. aeruginosa* MRC that only contains *mdnABC* genes, indicated that the precursor peptide and both ATP-grasp ligases are sufficient to produce correctly cyclised peptides. These results were confirmed by analyses of *E. coli* cells containing the minimal constructs with the *mdnABCD* genes from *M. aeruginosa* NIES298. Falsely processed but correctly cyclised variants of microviridin B were found in those cell extracts.

Unfortunately, further characterisations of the microviridin ligases *in vivo* by mutational analysis were not possible. Knock out mutations of one ATP-grasp ligase resulted in a complete loss of peptide expression, maybe due to rapid proteolysis of the precursor peptide in *E coli*.

Fast degradation of the small precursor protein was observed in other experiments as well. Using a lambda Red recombination system to knock out the ATP-grasp enzymes on the fosmids showed the same results (Weiz, personal communication), and initial approaches of heterologous expression of the MdnA precursor peptide in *E. coli* failed as well (data not shown). In order to prevent proteolysis in the cytosol, current investigations aim to express the precursor with a Dsb-fusion tag for export into the periplasm (Weiz, Dittmann; personal communications).

Instead of an *in vivo* analysis of the ATP-grasp ligases, characterisation was attempted *in vitro*. However, no activity was detectable using overexpressed his-tagged MdnB and C enzymes. Initial difficulties in the cloning process of the ATP-grasp ligase overexpression constructs indicated that higher amounts of these enzymes implicate severe side effects to the host. Selecting clones carrying the minimal construct in the pDrive vector was only possible when only one fourth of the usual antibiotic concentration was used. No clones were able to survive under higher selective pressure. Similar effects were observed in other laboratories (Anton Liaimer, personal communications). Even expression from the fosmid was only possible for a few hours, overnight expression cultures did only contain low amounts of microviridins if any at all. Maybe the *E. coli* host strain is only able to survive if the expressed ATP-grasp ligases are somehow inactivated, for example by misfolding. Ongoing experiments aim to express the microviridin ligases fused to a Dsb-tag for export and periplasmic folding, in order to avoid side effects and misfolding in *E. coli* (Weiz, Dittmann; personal communication).

Another approach to circumvent host problems would be the expression of the ligases by an *in vitro* transcription/translation system. Such a cell free system allows the production of proteins that, because of toxicity, generally pose problems in cell-based expression systems (Hoffmann *et al.*, 2004; Martin *et al.*, 2001).

Another possible reason of why the *in vitro* assay failed may lie in the absence of a required cofactor. Since heterologous expression of microviridins demonstrated that ATP-grasp ligases are active in cells of *E. coli*, activity assays with whole cell extracts from cells overexpressing MdnB and MdnC were performed. If there was any necessary cofactor, it should be present in the cell extracts. Unfortunately, soon after starting the experiment, the precursor peptide was degraded, in spite of the addition of various protease inhibitors. *In vitro* assays with ATP-grasp enzymes enriched by size exclusion chromatography failed as well.

Recently, MdnB and C homologues in *Planktothrix agardhii* were shown to be responsible for cyclisation of the microviridin K precursor *in vitro* (Philmus et al., 2008). In a similar approach applying the same experimental conditions as in this work, the microviridin ligases of *P. agardhii* were shown to be responsible for the ester and amide bonds in microviridin K without any additional cofactor except for magnesium and ATP. Philmus and colleagues were able to show that the ester bonds are first established by the MdnC homologue and then an MdnB homologue introduces the amide bond using the MdnC product as a substrate. Both enzymes were suggested to function by ATP activation of carboxylates to acylphosphate intermediates, which are then prone to attack by nucleophiles to yield amide and ester bonds.

According to the high inter-species similarity of the ATP-grasp ligases, it seems unlikely that the enzymes from *M. aeruginosa* need an additional cofactor, when those from *P. agardhii* do not. However, more studies are needed to understand the observed differences in the enzyme activities. An interesting approach would be an *in vitro* assay using the *P. agardhii* microviridin K precursor peptide as substrate for the microviridin ligases from *M. aeruginosa* NIES298. Philmus and colleagues were able to express the *P. agardhii* precursor without degradation problems in the same *E. coli* strain we chose for the microviridin B precursor peptide from *Microcystis aeruginosa* NIES298 (Philmus et al., 2008). Maybe the *P. agardhii* microviridin biosynthesis system is better suited for analysis in *E. coli*, due to sequence differences in the precursor peptides (fig. 37). However, further investigations are needed to verify these speculations.

```
MdnA    MAYPNDQQG-KALPFFARFLSVSKEESSIKSPSPEPTFGTTLKYPSDWEEY
MvdE    MSKNVKVSAPKAVPFFARFLAEQAVEANNSNSAP---YGNTMKYPSDWEEY
        *:   . .. **:*******:  . *:. ...:*    :*.*:*********
```

Fig. 37 Alignment of the microviridin B precursor from *Microcystis* aeruginosa NIES298 (MdnA) and the microviridin K precursor from *Planktothrix* agardhii CYA126/8 (MvdE). Microviridin encoding amino acids are shown in green.

How useful further charactisations of these biocatalyst can be, was demonstrated recently by a study where the substrate specifity and scope of the ester bond forming MdnC homologue from *P. agardhii* was analysed (Philmus *et al.*, 2009). The authors could show that the larger ring between threonine and aspartate is first established by an ester bond formation during microviridin K biosynthesis. The subsequently formed ester bond closes the smaller ring between serine and glutamate. Although the ring size for both ester bond formations is not flexible, at least for the first ring closure alanine substitutions in all positions not directly involved in the cross-linking of the microviridin K prepeptide were accepted, indicating a substrate tolerance that could be useful in producing novel natural compounds.

4.2.2 The N-acetyltransferase

Since all known microviridins are acetylated at their *N*-terminus, the presence of an *N*-acetyltransferase in the biosynthesis gene cluster is in perfect agreement with the structure. The Gcn5-related *N*-acetyltransferases (GNAT's) are an enormous superfamily of enzymes that comprises some 10,000 members in all kingdoms of life (Vetting *et al.*, 2005). GNAT's use acyl-CoAs to acylate their cognate substrates in a ternary complex. First members were identified as aminoglycoside acetyltransferases, which catalyse the regioselective acetylation of amino groups on aminoglycosides with antibiotic properties. Other substrates of members of the GNAT family are serotonin, glucosamine-6-phosphate, ribosomal proteins in bacteria and histones in eukaryotes.

Antibiotics, and the homoserine lactones (quorum sensing molecules from gram-negative bacteria) were shown to be acylated by *N*-acetyltransferases of the GNAT-family. Acetylation of microviridin K by MvdB, the MdnD homologue in *Planktothrix agardhii*, was shown *in vitro* (Philmus et al., 2008).

4.2.3 Which role does the ABC transporter play?

The putative ABC transporter present in the microviridin gene cluster in *M. aeruginosa* NIES298 shows similarities to ABC type exporters associated to non-ribosomally produced peptides. Although none of these transporters are characterised, knock-out mutations of the McyH-transporter from the microcystin NRPS gene cluster in *Microcystis aeruginosa* PCC7806 resulted in the loss of microcystin production and suggested an important role in the biosynthesis of these peptides (Pearson et al., 2004).

The first hint towards the functional role of the ABC type transporter in the biosynthesis of microviridins was given by the heterologous expression in *E. coli*. Whereas authentic microviridin B was produced by the strain carrying the *mdn* fosmid of *M. aeruginosa* NIES298, only false-processed variants are present in extracts of cells with the mdn fosmid from *M. aeruginosa* MRC or the minimal construct. Since both contain no *mdnE* gene, the ABC type transporter could be responsible for correct processing of the leader peptide. Recent studies revealed that the transporter indeed plays a role in the cleavage of the precursor. After coexpression of the NIES298 MdnE protein from an inducible expression vector along with the MRC *mdn* fosmid, correctly processed unacetylated microviridin J was detected. Coexpression of the *N*- acetyltransferase and the transporter with the fosmid resulted in production of authentic microviridin J (Weiz, Dittmann; unpublished data).

The exact mechanism of the cleavage remains elusive. In the case of bacteriocins and lantibiotics, ABC-tranporters containing an additional peptidase domain have been shown to cleave the *N*-terminal leader peptide from the respective precursor during transport through the inner cell membrane (Michiels et al., 2001).

Interestingly, the transporter encoded in the cryptic microviridin cluster from *Anabaena* sp. PCC7120, contains an additional peptidase domain at the N-terminus and the precursor peptide clearly shows consensus sequences of the double glycine motif (fig. 38), suggesting a bacteriocin-like processing in this cyanobacterium. In the case of *Microcystis* no such double glycine motif (fig. 38) and no peptidase domain on the transporter protein is detectable.

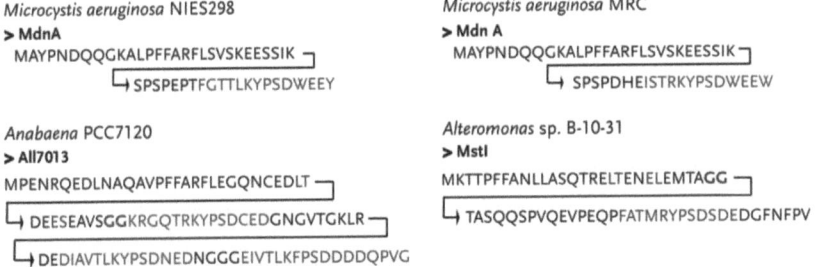

Fig. 38 Microviridin and marinostatin precursor peptides.
Peptide encoding sequences are indicated in green, double glycine consensus sequences in red.

So maybe the C-terminal part of the MdnE tranporter, which is absent in the Alr7014 microviridin associated transporter from *Anabaena* PCC7120 is somehow involved in the cleavage of the leader peptide using an unknown mechanism. Another possibility is that a protease also present in *E. coli* is responsible for the correct cleavage, and binding of the precursor peptide to the ABC transporter is needed for the process. The transporter could also be needed to maintain a structural complex, as it is thought for the microcystin biosynthesis (Pearson *et al.*, 2004).

The role of the ABC transporter in the transport of microviridin remains obscure. Sequence analysis of the microcystin transporter, McyH, suggested an export function for the protein and its role as a microcystin exporter was proposed (Pearson *et al.*, 2004). Although microcystin and other cyanobacterial peptides were found in the medium, usually about 90 per cent is detected intracellularly (Rapala *et al.*, 1997) (Ferreira, 2006).

HPLC analysis of growth medium supernatants of *Microcystis aeruginosa* NIES298 and MRC revealed no or only low amounts of microviridin, that were rather attributed to cell lysis than to extracellular transport (data not shown). In another study of *Microcystis* RST9501, the intracellular fraction amounted in most cases to over 80% of the total microviridin pool (Ferreira, 2006). However, exact conclusions are difficult to draw since the microviridin content inside cells shows a great variety (Keishi Ishida, personal communication; Ferreira, 2006).

4.2.4 Microviridins – another diverse family of cyanobacterial peptides

Similar to the microcyclamides, BLAST analysis of data base sequences and screening of *Microcystis* strains and field samples for microviridin precursor peptides revealed a high abundance of the biosynthesis genes. Microviridin precusor genes are not only present in *Microcystis*, *Planktothrix* and *Anabaena*, but in other cyanobacterial genera such as Nostoc, Nodularia and Cyanothece. One precursor was even found in the myxobacterium *Sorangium cellulosum*, and the marinostatin precursor from the marine bacteroidetes *Microscilla marina* clusters in this group as well. All precursor peptides contain the microviridin core motif TxK(Y/W/F)PSD, which has been shown to be important for the cyclisation reactions of the microviridin ligases (Philmus et al., 2009) and a consensus sequence of seven amino acids in the leader peptide PFFARFL. This motif could serve as a recognition sequence for the microviridin ligases. *In vitro* studies using microviridin precursor peptides with point mutations in this region as substrate for the microviridin ligases could answer that question. Screening bacterial genome sequences available in the databases for this motif could guide the identification of other compounds with related biosynthetic pathways.

The sporadic distribution in distantly related genera suggests horizontal gene transfer events responsible for the dispersal of microviridin biosynthesis genes, similar to the cyanobactin pathways. In contrast to cyanobactins, where a high variability was achieved by shuffling hypervariable peptide coding cassettes within the precursor peptides, most microviridin precursor peptides encode only one microviridin-like peptide, the exception being *Anabaena* sp. PCC7120, in which the precursor encodes for three peptides. If more than one microviridin encoding sequence is detectable, it is encoded on an additional precursor protein. *Planktothrix agardhii* CYA126/8 encodes two precursor and *Cyanothece* sp. PCC7822 even eight. Therefore, the microviridin family of peptides is a new example for a natural peptide library.

This is confirmed by the remarkable diversity observed in *Microcystis* laboratory strains and field samples. Sceening 18 strains sampled from all over the world (tab. 2, chapter 2.1.1) revealed six novel putative microviridins. Seven additional new microviridins were detected in the field samples. Alignment of all obtained microviridin sequences showed, to some extent, identical leader peptides and the conserved core motif, only the first three to four and the last amino acid positions are flexible. *In vitro* analysis of the microviridin ligases of *P. agardhii* CYA126/8 with manipulated precursor peptides revealed that the cross-linking of microviridins is strictly ordered and the ring size is not flexible (Philmus et al., 2009). Although alanine substitutions in all amino acid positions not directly involved in cross-linking were possible to detect the first ester bond formation between threonine and aspartate, the conserved core motif seems necessary to obtain the typical tricyclic structure. These results are supported by the fact, that every detected naturally occuring microviridin precursor contains the core motif, indicating that the tricyclic architecture is important for the bioactivity of microviridins.

It is noteworthy, that more than half of the sequenced clones derived from the field samples contained the peptide encoding sequence YNVTLKYPSDWEEF, suggesting the highest abundance of these precursor genes in the analysed natural habitats. No microviridin with such amino acid sequence is known. Therefore, structural elucidation and bioactivity assays with this variant would be most interesting. Moreover, information about their biological activity could provide key insights into the ecological role of microviridins in the field.

4.2.5 Possible applications in bioengineering

Screening the biosynthesis genes and especially the precursor proteins of ribosomal peptides can guide the discovery of new compounds. As already shown for the microcyclamides, the microviridins gave another example for a successful genomic mining approach. The microviridin peptide from strain NIES100 is already identified and structurally elucidated by Keishi Ishida and colleagues at the HKI in Jena (unpublished data).

However, successful heterologous expression of different members of this peptide family in *E. coli* provides the basis for many further studies. Cyanobacterial species such as *Microcystis* are slow-growing bacteria and hardly amenable to genetic manipulations. They produce a variety of different peptides and peptide families, which could be subjected to complex regulatory mechanisms, as shown for the microviridin biosynthesis genes in *M. aeruginosa* NIES298 and MRC. Therefore, transfer of metabolite pathways into a suitable heterologous host could facilitate the identification and structural elucidation of novel peptides, even if the biosynthetic genes are silent in their natural producers.

This approach has been demonstrated for a silent type III polyketide synthase of the myxobacterium *Sorangium cellulosum* Soce56. The corresponding product flaviolin was identified by heterologous expression in three different pseudomonads (Gross *et al.*, 2006). A number of NRPS and PKS metabolites from slow-growing and difficult to handle myxobacteria and actinomycetes have been characterised by transfer of the biosynthetic genes to more amenable hosts (Wenzel, Müller, 2005). Ribosomal peptides were also analysed by this method. Patellamides and related peptides were expressed in *E. coli*, which enabled the identification and characterisation of the biosynthetic genes from a non-culturable *Prochloron* strain (Schmidt *et al.*, 2005).

The example of patellamides and microviridins revealed that *E. coli* seems a suitable host for the production of cyanobacterial ribosomal metabolites, whereas pseudomonads, bacilli or fast growing *Streptomyces* species are more suitable hosts for the heterologous expression of NRPS and PKS from myxobacteria and slow growing streptomycetes (Wenzel, Müller, 2005). Although ATP-grasp ligases seem to cause some side effects, recognition of the cyanobacterial promotors, a similar codon usage and very fast growth rates are promising properties of *E. coli* as a host strain. Microviridin biosynthesis genes from other *Microcystis* strains could be transferred in *E. coli* and expressed metabolites could be easily identified by comparison of metabolic profiles to the wild-type host. As proof of principle, the microviridin from *Microcystis* NIES843 is currently under investigation. Heterologous expression of the peptide in *E. coli* was already successful and, in cooperation with Keishi Ishida and Christian Hertweck from Jena, structural elucidation is in progress (Annika Weiz, personal communications). Furthermore, direct expression of microviridins from fosmids provide the opportunity of screening metagenomic libraries for the production of novel variants and circumvent the operose cultivation of laboratory strains.

The most promising tool provided by heterologous expression of microviridins, is the production of large peptide libraries through biosynthetic engineering. Ongoing experiments tried to establish an easily manipulable and stable expression system in *E. coli* (Weiz, Dittmann; personal communications) and fast growing cyanobacteria such as *Synechocystis* (Liaimer, personal communications). Point mutations in every possible amino acid in the precursor peptide including the leader peptide or the replacement of whole regions, would not only provide insights into biosynthetic mechanisms but broaden the natural variety of microviridins.

Philmus and colleagues already showed that at least the MdnC-homologous microviridin ligase from *P. agardhii* CYA126/8 possess some limited substrate tolerance to perform ester bond formations (Philmus *et al.*, 2009). Structure refinement by protein crystallisation and detailed computational and comparative analyses of the microviridin ligases may facilitate effective engineering to ultimately modulate the substrate specificity and introduce cross-links into microviridin-unrelated metabolites or even proteins to obtain novel bioactivities.

4.2.6 Functional hypotheses for microviridins

Microviridins are discussed as feeding deterrent against grazing *Daphnia*, although only one peptide of this family, microviridin J, was shown to be toxic for *Daphnia* (Rohrlack et al., 2003). However, the huge variety of these metabolites in cyanobacteria suggests further functions for their producers. Expression analyses of microviridin biosynthesis genes were to obtain first hints towards possible regulatory pathways. In contrast to microcyclamide from *M. aeruginosa* NIES298, which is transcribed in a more or less constitutive pattern independently from growth phases and light intensities, the microviridin ligase expression was shown to be strongly regulated.

Transcription and expression data from the peptides correspond to observed peptide abundances in the cells. Whereas microcyclamide was always detectable in *M. aeruginosa* NIES298, microviridin B content varied strongly (data not shown). Unfortunately, no precise conditions could be assigned to the upregulation of microviridin biosynthesis genes. These results are supported by a former study with *Microcystis* RST9501. Extra- and intracellular microviridin concentrations were analysed and very variable amounts of microviridin, independent from temperature and light intensity, were observed (Ferreira, 2006).

The generation of an antibody against one of the microviridin ligases would provide the opportunity for further expression analyses that were beyond the scope of this study. Nutrient limitations or CO_2 concentration could influence the expression of biosynthesis genes as for example was shown for microcystins (Rapala et al., 1997). Co-cultivations of other organisms could induce the production of microviriridins. Recently, it has been shown that intimate interaction with an actinobacterium triggers the production of a polyketide in the fungus *Aspergillus nidulans* (Schroeckh et al., 2009). As microviridins are discussed as feeding deterrent, co-cultivation with *Daphnia* species or addition of *Daphnia* extracts to *Microcystis* growth media could be analysed.

Since the generated anti-MdnB antibody binds to the ligase encoded in *Anabaena* as well, studies in other cyanobacteria could provide further informations. Furthermore, the *Anabaena* strain PCC7120 is, in contrast to *Microcystis aeruginosa* NIES298 and MRC, genetically manipulable (Fiedler et al., 1998). Although no microviridin-like peptides were detected in this strain, promotor fusions to reporters such as the GFP-protein could facilitate the search for regulatory mechanisms. Microviridin biosynthesis genes were also characterised in *Planktothrix agardhii* CYA126/8 (Philmus et al., 2008), a cyanobacterial strain that is reliably amenable to genetic manipulations (Christiansen et al., 2003; Ishida et al., 2007). Both strains could be used for mutational studies in microviridin biosynthesis *in vivo*.

Another hint at possible functions of microviridin was provided by the immunofluorescence microscopy using the anti-MdnB antibody. Specific signals were detected surrounding the septum of dividing cells. Unfortunately, not all cells in that status showed those signals. Furthermore, in the vicinity of microviridin biosynthesis gene clusters in *Microcystis aeruginosa* NIES298 and NIES843 genes coding for FtsW and MinD homologues were found. Both proteins are known to be involved in cell cycle regulation in various bacterial genera (Donachie, 1993; Leonard et al., 2005). However, no cell-cycle associated genes were detectable in the closer vicinity of microviridin pathways of other cyanobacteria. A possible intracellular function of a secondary metabolite has recently been shown for microcystins. Microcystin was shown to influence the abundance of several proteins, its covalent binding to RubisCO and other proteins involved in the Calvin cycle and its physiological function in light and redox-regulated processes within the cell was discussed (Zilliges, 2007).

However, much more analysis and informations are needed to verify or disprove these speculations. Identification of the biosynthetic mechanisms and initial analysis of regulatory pathways could only provide a basis for further investigations to point out possible functions of microviridins.

4.3 General consideration about possible functions of cyanobacterial secondary metabolites

One striking feature of secondary metabolites from cyanobacteria emerging from the literature and from data obtained in this study is the huge variability of peptides and peptide families and their global distribution. The best characterised peptide family from cyanobacteria, the microcystins, comprise more than 60 known peptides isolated from cyanobacteria from various climatic zones such as tropical lakes, coastal waters and Antarctic ice (Welker, von Döhren, 2006). Furthermore, multiple peptides and peptide classes are co-produced by individual strains. Strains from genera such as *Nostoc*, *Anabaena* and *Planktothrix* usually contain up to four peptide classes. The combination of peptides allows the discrimination of morphologically undistinguishable strains as chemotypes, with typical peptide fingerprints (Fastner *et al.*, 2001; Welker *et al.*, 2004). *Microcystis aeruginosa* NIES298 for example produces, additionally to microviridin and microcyclamide, microcystin and aeruginosin (Fukuta *et al.*, 2004). The co-occurence of different chemotypes in natural habitats results in an even higher variability of peptides (Fastner *et al.*, 2001; Welker *et al.*, 2004).

Various hypotheses have been formulated regarding the function of microcystins and other peptides in the physiology and ecology of cyanobacteria. One of the most common proposes their role as feeding deterrent from *Daphnia* and other grazing zooplankton species. Toxicity to invertebrate animals results mostly from inhibitory effects on different enzymes of the grazers. Microcystin for example inhibits the protein phosphatases from *Daphnia* (Rohrlack *et al.*, 1999), whereas microviridin J was shown to inhibit the daphnid trypsin-like proteases (Rohrlack *et al.*, 2003). However, not all cyanobacterial peptides possess inhibitory activities. Furthermore, colony formation or filament formation has been recognised as the most important factor in grazing resistance of cyanobacteria (Hansson *et al.*, 1998; Welker, von Döhren, 2006).

Another fact arguing against a primary role of cyanobacterial peptides in grazing protection is the early evolution of the biosynthesis genes. Phylogenetic analyses of the microcystin biosynthesis genes in *Microcystis*, *Anabaena* and *Planktothrix* revealed, that these genes are a very ancient part of the cyanobacterial metabolism, which evolved long before higher plants or animals existed (Rantala et al., 2004).

Only few examples are known where distinct physiological roles for cyanobacterial secondary metabolites could be identified. In *Anabaena* sp. PCC7120 a small peptide pheromone was shown to inhibit heterocyst formation in cells adjacent to existing heterocysts (Golden, Yoon, 2003; Zhang et al., 2006). As mentioned before, the microcystins were suggested to be involved in cell-cell recognition of *Microcystis* species (Kehr et al., 2006) and in redox-regulation within the cells (Zilliges, 2007). A very recent example revealed an important role of a polyketide in cellular differentiation of the symbiotic cyanobacterium *Nostoc punctiforme* (Liaimer and Dittmann, submitted). Insertional mutagenesis of the polyketide biosynthesis genes led to the accumulation of short filaments in the hormogonia and primordial states, two exclusive cell types, that are important in symbiosis mechanism between the cyanobacteria and plants.

These examples suggest that the huge variety of peptides produced via non-ribosomal and ribosomal pathways could have evolved and maybe is still evolving towards a similar variety of functional roles of these compounds in cyanobacterial lifestyles. However, more data are needed to draw well-founded conclusions. Combined bioinformatic approaches, molecular analyses as well as field studies can provide insights into the production of natural compounds not only to use their variety in drug development, but also to understand their diverse occurence in the environment.

5 References

Angermayr SA, Hellingwerf KJ, Lindblad P, de Mattos MJ (2009) Energy biotechnology with cyanobacteria. *Curr Opin Biotechnol* **20**, 257-263.

Arulanantham H, Kershaw NJ, Hewitson KS, et al. (2006) ORF17 from the clavulanic acid biosynthesis gene cluster catalyzes the ATP-dependent formation of N-glycyl-clavaminic acid. *J Biol Chem* **281**, 279-287.

Baba T, Schneewind O (1998) Instruments of microbial warfare: bacteriocin synthesis, toxicity and immunity. *Trends Microbiol* **6**, 66-71.

Baker DD, Chu M, Oza U, Rajgarhia V (2007) The value of natural products to future pharmaceutical discovery. *Nat Prod Rep* **24**, 1225-1244.

Banker R, Carmeli S (1998) Tenuecyclamides A-D, cyclic hexapeptides from the cyanobacterium *Nostoc spongiaeforme var. tenue*. *J Nat Prod* **61**, 1248-1251.

Bentley SD, Chater KF, Cerdeno-Tarraga AM, et al. (2002) Complete genome sequence of the model actinomycete *Streptomyces coelicolor* A3(2). *Nature* **417**, 141-147.

Berg H, Ziegler K, Piotukh K, et al. (2000) Biosynthesis of the cyanobacterial reserve polymer multi-L-arginyl-poly-L-aspartic acid (cyanophycin): mechanism of the cyanophycin synthetase reaction studied with synthetic primers. *Eur J Biochem* **267**, 5561-5570.

Berry JP, Gantar M, Perez MH, Berry G, Noriega FG (2008) Cyanobacterial toxins as allelochemicals with potential applications as algaecides, herbicides and insecticides. *Mar Drugs* **6**, 117-146.

Binder M, Tamm C (1973) The cytochalasans: a new class of biologically active microbial metabolites. *Angew Chem Int Ed Engl* **12**, 370-380.

Brakhage AA (1997) Molecular regulation of penicillin biosynthesis in *Aspergillus* (*Emericella*) *nidulans*. *FEMS Microbiol Lett* **148**, 1-10.

Breukink E (2006) A lesson in efficient killing from two-component lantibiotics. *Mol Microbiol* **61**, 271-273.

Cadel-Six S, Dauga C, Castets AM, et al. (2008) Halogenase genes in nonribosomal peptide synthetase gene clusters of *Microcystis* (cyanobacteria): sporadic distribution and evolution. *Mol Biol Evol* **25**, 2031-2041.

Challis GL, Hopwood DA (2003) Synergy and contingency as driving forces for the evolution of multiple secondary metabolite production by *Streptomyces* species. *Proc Natl Acad Sci U S A* **100 Suppl 2**, 14555-14561.

Chorus I, Falconer IR, Salas HJ, Bartram J (2000) Health risks caused by freshwater cyanobacteria in recreational waters. *J Toxicol Environ Health B Crit Rev* **3**, 323-347.

Christiansen G, Fastner J, Erhard M, Börner T, Dittmann E (2003) Microcystin biosynthesis in *Planktothrix*: genes, evolution, and manipulation. *J Bacteriol* **185**, 564-572.

Czarnecki O, Henning M, Lippert I, Welker M (2006) Identification of peptide metabolites of *Microcystis* (Cyanobacteria) that inhibit trypsin-like activity in planktonic herbivorous *Daphnia* (Cladocera). *Environ Microbiol* 8, 77-87.

Davies J (2006) Are antibiotics naturally antibiotics? *J Ind Microbiol Biotechnol* 33, 496-499.

Degnan BM, Hawkins CJ, Lavin MF, et al. (1989) Novel cytotoxic compounds from the ascidian *Lissoclinum bistratum*. *J Med Chem* 32, 1354-1359.

Delves-Broughton J, Blackburn P, Evans RJ, Hugenholtz J (1996) Applications of the bacteriocin, nisin. *Antonie Van Leeuwenhoek* 69, 193-202.

Demain AL, Fang A (2000) The natural functions of secondary metabolites. *Adv Biochem Eng Biotechnol* 69, 1-39.

Des Marais DJ (1991) Microbial mats, stromatolites and the rise of oxygen in the Precambrian atmosphere. *Glob Planet Change* 97, 93-96.

Dobrindt U, Hochhut B, Hentschel U, Hacker J (2004) Genomic islands in pathogenic and environmental microorganisms. *Nat Rev Microbiol* 2, 414-424.

Doekel S, Eppelmann K, Marahiel MA (2002) Heterologous expression of nonribosomal peptide synthetases in B. subtilis: construction of a bi-functional B subtilis/E. coli shuttle vector system. *FEMS Microbiol Lett* 216, 185-191.

Donachie WD (1993) The cell cycle of *Escherichia coli*. *Annu Rev Microbiol* 47, 199-230.

Donadio S, Monciardini P, Sosio M (2007) Polyketide synthases and nonribosomal peptide synthetases: the emerging view from bacterial genomics. *Nat Prod Rep* 24, 1073-1109.

Donia MS, Hathaway BJ, Sudek S, et al. (2006) Natural combinatorial peptide libraries in cyanobacterial symbionts of marine ascidians. *Nat Chem Biol* 2, 729-735.

Donia MS, Ravel J, Schmidt EW (2008) A global assembly line for cyanobactins. *Nat Chem Biol* 4, 341-343.

Fastner J, Erhard M, von Döhren H (2001) Determination of oligopeptide diversity within a natural population of *Microcystis* spp. (cyanobacteria) by typing single colonies by matrix-assisted laser desorption ionization-time of flight mass spectrometry. *Appl Environ Microbiol* 67, 5069-5076.

Ferreira AH (2006) *Peptides In Cyanobacteria Under Different Environmental Conditions* Text. Doctoral Thesis. Technische Universität Berlin.

Fiedler G, Arnold M, Hannus S, Maldener I (1998) The DevBCA exporter is essential for envelope formation in heterocysts of the cyanobacterium *Anabaena* sp. strain PCC 7120. *Mol Microbiol* 27, 1193-1202.

Field CB, Behrenfeld MJ, Randerson JT, Falkowski P (1998) Primary production of the biosphere: integrating terrestrial and oceanic components. *Science* 281, 237-240.

Finking R, Marahiel MA (2004) Biosynthesis of nonribosomal peptides. *Annu Rev Microbiol* 58, 453-488.

Firn RD, Jones CG (2000) The evolution of secondary metabolism - a unifying model. *Mol Microbiol* **37**, 989-994.

Firn RD, Jones CG (2003) Natural products--a simple model to explain chemical diversity. *Nat Prod Rep* **20**, 382-391.

Firn RD, Jones CG (2006) Do we need a new hypothesis to explain plant VOC emissions? *Trends Plant Sci* **11**, 112-113; author reply 113-114.

Fischbach MA, Walsh CT (2006) Assembly-line enzymology for polyketide and nonribosomal Peptide antibiotics: logic, machinery, and mechanisms. *Chem Rev* **106**, 3468-3496.

Fleming A (1929) On the Antibacterial Action of Cultures of a *Penicillium*, with Special Reference to their Use in the Isolation of B. *Influenzae*. *British Journal of Experimental Pathology* **10**, 226-236

Franche C, Damerval T (1988) Tests on nif Probes and DNA Hybridizations. In: *Methods in Enzymology* (eds. Packer L, Glazer AN). Academic Press, Inc., San Diego.

Frangeul L, Quillardet P, Castets AM, *et al.* (2008) Highly plastic genome of *Microcystis aeruginosa* PCC 7806, a ubiquitous toxic freshwater cyanobacterium. *BMC Genomics* **9**, 274.

Frickey T, Lupas A (2004) CLANS: a Java application for visualizing protein families based on pairwise similarity. *Bioinformatics* **20**, 3702-3704.

Fukuta Y, Ohshima T, Gnanadesikan V, *et al.* (2004) Enantioselective syntheses and biological studies of aeruginosin 298-A and its analogs: application of catalytic asymmetric phase-transfer reaction. *Proc Natl Acad Sci U S A* **101**, 5433-5438.

Gehring AM, Mori I, Walsh CT (1998) Reconstitution and characterization of the *Escherichia coli* enterobactin synthetase from EntB, EntE, and EntF. *Biochemistry* **37**, 2648-2659.

Gerwick WH, Tan LT, Sitachitta N (2001) Nitrogen-containing metabolites from marine cyanobacteria. *Alkaloids Chem Biol* **57**, 75-184.

Giovannoni SJ, Turner S, Olsen GJ, *et al.* (1988) Evolutionary relationships among cyanobacteria and green chloroplasts. *J Bacteriol* **170**, 3584-3592.

Golden JW, Yoon HS (2003) Heterocyst development in *Anabaena*. *Curr Opin Microbiol* **6**, 557-563.

Gross F, Luniak N, Perlova O, *et al.* (2006) Bacterial type III polyketide synthases: phylogenetic analysis and potential for the production of novel secondary metabolites by heterologous expression in pseudomonads. *Arch Microbiol* **185**, 28-38.

Gross H (2007) Strategies to unravel the function of orphan biosynthesis pathways: recent examples and future prospects. *Appl Microbiol Biotechnol* **75**, 267-277.

Guljamow A, Jenke-Kodama H, Saumweber H, *et al.* (2007) Horizontal gene transfer of two cytoskeletal elements from a eukaryote to a cyanobacterium. *Curr Biol* **17**, R757-759.

Hahn M, Stachelhaus T (2004) Selective interaction between nonribosomal peptide synthetases is facilitated by short communication-mediating domains. *Proc Natl Acad Sci U S A* **101**, 15585-15590.

Hall TA (1999) BioEdit: a user-friendly biological sequence alignment editor and analysis program for Windows 95/98/NT. *Nucl Acids Symp Ser* **41**, 95 - 98.

Hallen HE, Luo H, Scott-Craig JS, Walton JD (2007) Gene family encoding the major toxins of lethal *Amanita* mushrooms. *Proc Natl Acad Sci U S A* **104**, 19097-19101.

Hansson LA, Bergman E, Cronberg G (1998) Size structure and succession in phytoplankton communities: the impact of interactions between herbivory and predation. *Oikos* **81**, 337-345.

Hartmann T (2004) Plant-derived secondary metabolites as defensive chemicals in herbivorous insects: a case study in chemical ecology. *Planta* **219**, 1-4.

Hawkins CJ, Lavin MF, Marshall KA, van den Brenk AL, Watters DJ (1990) Structure-activity relationships of the lissoclinamides: cytotoxic cyclic peptides from the ascidian *Lissoclinum patella*. *J Med Chem* **33**, 1634-1638.

Hennings H, Blumberg PM, Pettit GR, et al. (1987) Bryostatin 1, an activator of protein kinase C, inhibits tumor promotion by phorbol esters in SENCAR mouse skin. *Carcinogenesis* **8**, 1343-1346.

Hertweck C (2009) The biosynthetic logic of polyketide diversity. *Angew Chem Int Ed Engl* **48**, 4688-4716.

Hoffmann D, Hevel JM, Moore RE, Moore BS (2003) Sequence analysis and biochemical characterization of the nostopeptolide A biosynthetic gene cluster from *Nostoc* sp. GSV224. *Gene* **311**, 171-180.

Hoffmann M, Nemetz C, Madin K, Buchberger B (2004) Rapid translation system: a novel cell-free way from gene to protein. *Biotechnol Annu Rev* **10**, 1-30.

Hoiczyk E, Hansel A (2000) Cyanobacterial cell walls: news from an unusual prokaryotic envelope. *J Bacteriol* **182**, 1191-1199.

Huelsenbeck JP, Ronquist F (2001) MRBAYES: Bayesian inference of phylogenetic trees. *Bioinformatics* **17**, 754-755.

Ichinose K, Bedford DJ, Tornus D, et al. (1998) The granaticin biosynthetic gene cluster of *Streptomyces violaceoruber* Tu22: sequence analysis and expression in a heterologous host. *Chem Biol* **5**, 647-659.

Ishida K, Christiansen G, Yoshida WY, et al. (2007) Biosynthesis and structure of aeruginoside 126A and 126B, cyanobacterial peptide glycosides bearing a 2-carboxy-6-hydroxyoctahydroindole moiety. *Chem Biol* **14**, 565-576.

Ishida K, Nakagawa H, Murakami M (2000) Microcyclamide, a cytotoxic cyclic hexapeptide from the cyanobacterium *Microcystis* aeruginosa. *J Nat Prod* **63**, 1315-1317.

Ishida K, Welker M, Christiansen G, et al. (2009) Plasticity and evolution of aeruginosin biosynthesis in cyanobacteria. *Appl Environ Microbiol* **75**, 2017-2026.

Ishitsuka MO, Kusumi T, Kakisawa H, Kaya K, Watanabe MM (1990) Microviridin - a Novel Tricyclic Depsipeptide from the Toxic Cyanobacterium *Microcystis viridis*. *Journal of the American Chemical Society* 112, 8180-8182.

Jack RW, Jung G (2000) Lantibiotics and microcins: polypeptides with unusual chemical diversity. *Curr Opin Chem Biol* 4, 310-317.

Jenke-Kodama H, Dittmann E (2009a) Bioinformatic perspectives on NRPS/PKS megasynthases: Advances and challenges. *Nat Prod Rep* 26, 874-883.

Jenke-Kodama H, Dittmann E (2009b) Evolution of metabolic diversity: Insights from microbial polyketide synthases. *Phytochemistry*.

Jensen PR, Fenical W (1994) Strategies for the discovery of secondary metabolites from marine bacteria: ecological perspectives. *Annu Rev Microbiol* 48, 559-584.

Jones DT, Taylor WR, Thornton JM (1992) The rapid generation of mutation data matrices from protein sequences. *Comput Appl Biosci* 8, 275-282.

Jüttner F, Todorova AK, Walch N, von Philipsborn W (2001) Nostocyclamide M: a cyanobacterial cyclic peptide with allelopathic activity from *Nostoc* 31. *Phytochemistry* 57, 613-619.

Kaebernick M, Rohrlack T, Christoffersen K, Neilan BA (2001) A spontaneous mutant of microcystin biosynthesis: genetic characterization and effect on *Daphnia*. *Environ Microbiol* 3, 669-679.

Kaneko T, Nakajima N, Okamoto S, et al. (2007) Complete genomic structure of the bloom-forming toxic cyanobacterium *Microcystis* aeruginosa NIES-843. *DNA Res* 14, 247-256.

Kehr JC, Zilliges Y, Springer A, et al. (2006) A mannan binding lectin is involved in cell-cell attachment in a toxic strain of *Microcystis aeruginosa*. *Mol Microbiol* 59, 893-906.

Kleerebezem M (2004) Quorum sensing control of lantibiotic production; nisin and subtilin autoregulate their own biosynthesis. *Peptides* 25, 1405-1414.

Kleerebezem M, Quadri LE, Kuipers OP, de Vos WM (1997) Quorum sensing by peptide pheromones and two-component signal-transduction systems in Gram-positive bacteria. *Mol Microbiol* 24, 895-904.

König GM, Kehraus S, Seibert SF, Abdel-Lateff A, Müller D (2006) Natural products from marine organisms and their associated microbes. *Chembiochem* 7, 229-238.

Laemmli UK (1970) Cleavage of structural proteins during the assembly of the head of bacteriophage T4. *Nature* 227, 680-685.

Larget B, Simon DL (1999) Markov chain Monte Carlo algorithms for the Bayesian analysis of phylogenetic trees. *Molecular Biology and Evolution* 16, 750-759.

Lee J, McIntosh J, Hathaway BJ, Schmidt EW (2009) Using marine natural products to discover a protease that catalyzes peptide macrocyclization of diverse substrates. *J Am Chem Soc* 131, 2122-2124.

Lee SW, Mitchell DA, Markley AL, et al. (2008) Discovery of a widely distributed toxin biosynthetic gene cluster. Proc Natl Acad Sci U S A **105**, 5879-5884.

Leikoski N, Fewer DP, Sivonen K (2009) Widespread occurrence and lateral transfer of the cyanobactin biosynthesis gene cluster in cyanobacteria. Appl Environ Microbiol **75**, 853-857.

Leonard TA, Moller-Jensen J, Lowe J (2005) Towards understanding the molecular basis of bacterial DNA segregation. Philos Trans R Soc Lond B Biol Sci **360**, 523-535.

Li JW, Vederas JC (2009) Drug discovery and natural products: end of an era or an endless frontier? Science **325**, 161-165.

Li YM, Milne JC, Madison LL, Kolter R, Walsh CT (1996) From peptide precursors to oxazole and thiazole-containing peptide antibiotics: microcin B17 synthase. Science **274**, 1188-1193.

Linington RG, Clark BR, Trimble EE, et al. (2009) Antimalarial peptides from marine cyanobacteria: isolation and structural elucidation of gallinamide A. J Nat Prod **72**, 14-17.

Linington RG, Gonzalez J, Urena LD, et al. (2007) Venturamides A and B: antimalarial constituents of the panamanian marine Cyanobacterium Oscillatoria sp. J Nat Prod **70**, 397-401.

Long PF, Dunlap WC, Battershill CN, Jaspars M (2005) Shotgun cloning and heterologous expression of the patellamide gene cluster as a strategy to achieving sustained metabolite production. Chembiochem **6**, 1760-1765.

Macdonald KD, Holt G (1976) Genetics of biosynthesis and overproduction of penicillin. Sci Prog **63**, 547-573.

Maplestone RA, Stone MJ, Williams DH (1992) The evolutionary role of secondary metabolites--a review. Gene **115**, 151-157.

Marahiel MA, Stachelhaus T, Mootz HD (1997) Modular Peptide Synthetases Involved in Nonribosomal Peptide Synthesis. Chem Rev **97**, 2651-2674.

Martin GA, Kawaguchi R, Lam Y, et al. (2001) High-yield, in vitro protein expression using a continuous-exchange, coupled transcription/ translation system. Biotechniques **31**, 948-950, 952-943.

McIntosh JA, Donia MS, Schmidt EW (2009) Ribosomal peptide natural products: bridging the ribosomal and nonribosomal worlds. Natural Product Reports **26**, 537-559.

Meiser P, Bode HB, Müller R (2006) The unique DKxanthene secondary metabolite family from the myxobacterium Myxococcus xanthus is required for developmental sporulation. Proc Natl Acad Sci U S A **103**, 19128-19133.

Menzella HG, Reeves CD (2007) Combinatorial biosynthesis for drug development. Curr Opin Microbiol **10**, 238-245.

Michiels J, Dirix G, Vanderleyden J, Xi C (2001) Processing and export of peptide pheromones and bacteriocins in Gram-negative bacteria. Trends Microbiol **9**, 164-168.

Milne JC, Eliot AC, Kelleher NL, Walsh CT (1998) ATP/GTP hydrolysis is required for oxazole and thiazole biosynthesis in the peptide antibiotic microcin B17. *Biochemistry* **37**, 13250-13261.

Miyamoto K, Tsujibo H, Hikita Y, et al. (1998) Cloning and nucleotide sequence of the gene encoding a serine proteinase inhibitor named marinostatin from a marine bacterium, Alteromonas sp. strain B-10-31. *Biosci Biotechnol Biochem* **62**, 2446-2449.

Molinski TF, Dalisay DS, Lievens SL, Saludes JP (2009) Drug development from marine natural products. *Nat Rev Drug Discov* **8**, 69-85.

Montoya JP, Holl CM, Zehr JP, et al. (2004) High rates of N2 fixation by unicellular diazotrophs in the oligotrophic Pacific Ocean. *Nature* **430**, 1027-1032.

Mori T, Gustafson KR, Pannell LK, et al. (1998) Recombinant production of cyanovirin-N, a potent human immunodeficiency virus-inactivating protein derived from a cultured cyanobacterium. *Protein Expr Purif* **12**, 151-158.

Mur L, Skulberg O, Utkilen H (1999) Cyanobacteria in the environment. In: *Toxic Cyanobacteria in Water: A guide to their public health consequences, monitoring and management* (eds. Chorus I, Bartram J). E & FN Spon on behalf of WHO, London and New York

Murakami M, Sun Q, Ishida K, et al. (1997) Microviridins, elastase inhibitors from the cyanobacterium Nostoc minutum (NIES-26). *Phytochemistry* **45**, 1197-1202.

Namikoshi M, Rinehart KL (1996) Bioactive compounds produced by cyanobacteria. *Journal of Industrial Microbiology & Biotechnology* **17**, 373-384.

Nishizawa A, Arshad AB, Nishizawa T, et al. (2007) Cloning and characterization of a new hetero-gene cluster of nonribosomal peptide synthetase and polyketide synthase from the cyanobacterium *Microcystis* aeruginosa K-139. *J Gen Appl Microbiol* **53**, 17-27.

Ogino J, Moore RE, Patterson GM, Smith CD (1996) Dendroamides, new cyclic hexapeptides from a blue-green alga. Multidrug-resistance reversing activity of dendroamide A. *J Nat Prod* **59**, 581-586.

Okino T, Matsuda H, Murakami M, Yamaguchi K (1993) Microginin, an Angiotensin-Converting Enzyme-Inhibitor from the Blue-Green-Alga *Microcystis*-Aeruginosa. *Tetrahedron Letters* **34**, 501-504.

Okino T, Matsuda H, Murakami M, Yamaguchi K (1995) New Microviridins, Elastase Inhibitors from the Blue-Green-Alga *Microcystis-Aeruginosa*. *Tetrahedron* **51**, 10679-10686.

Olivera BM (2006) Conus peptides: biodiversity-based discovery and exogenomics. *J Biol Chem* **281**, 31173-31177.

Onaka H, Nakaho M, Hayashi K, Igarashi Y, Furumai T (2005) Cloning and characterization of the goadsporin biosynthetic gene cluster from *Streptomyces* sp. TP-A0584. *Microbiology* **151**, 3923-3933.

Otsuka S, Suda S, Li R, Matsumoto S, Watanabe MM (2000) Morphological variability of colonies of *Microcystis* morphospecies in culture. *J Gen Appl Microbiol* **46**, 39-50.

Paerl HW, Huisman J (2008) Climate. Blooms like it hot. *Science* **320**, 57-58.

Partida-Martinez LP, Hertweck C (2005) Pathogenic fungus harbours endosymbiotic bacteria for toxin production. *Nature* **437**, 884-888.

Partida-Martinez LP, Hertweck C (2007) A gene cluster encoding rhizoxin biosynthesis in "*Burkholderia rhizoxina*", the bacterial endosymbiont of the fungus *Rhizopus* microsporus. *Chembiochem* **8**, 41-45.

Paul VJ, Arthur KE, Ritson-Williams R, Ross C, Sharp K (2007) Chemical defenses: from compounds to communities. *Biol Bull* **213**, 226-251.

Pearson LA, Hisbergues M, Börner T, Dittmann E, Neilan BA (2004) Inactivation of an ABC transporter gene, mcyH, results in loss of microcystin production in the cyanobacterium *Microcystis* aeruginosa PCC 7806. *Appl Environ Microbiol* **70**, 6370-6378.

Philmus B, Christiansen G, Yoshida WY, Hemscheidt TK (2008) Post-translational modification in microviridin biosynthesis. *Chembiochem* **9**, 3066-3073.

Philmus B, Guerrette JP, Hemscheidt TK (2009) Substrate specificity and scope of MvdD, a GRASP-like ligase from the microviridin biosynthetic gene cluster. *ACS Chem Biol* **4**, 429-434.

Piel J (2009) Metabolites from symbiotic bacteria. *Nat Prod Rep* **26**, 338-362.

Portmann C, Blom JF, Gademann K, Jüttner F (2008a) Aerucyclamides A and B: isolation and synthesis of toxic ribosomal heterocyclic peptides from the cyanobacterium *Microcystis* aeruginosa PCC 7806. *J Nat Prod* **71**, 1193-1196.

Portmann C, Blom JF, Kaiser M, *et al.* (2008b) Isolation of Aerucyclamides C and D and Structure Revision of Microcyclamide 7806A: Heterocyclic Ribosomal Peptides from *Microcystis* aeruginosa PCC 7806 and Their Antiparasite Evaluation. *J Nat Prod*.

Potterat O, Hamburger M (2008) Drug discovery and development with plant-derived compounds. *Prog Drug Res* **65**, 45, 47-118.

Rai A (1990) Handbook of symbiotic cyanobacteria, p. 253 p. CRC Press, Boca Raton.

Rantala A, Fewer DP, Hisbergues M, *et al.* (2004) Phylogenetic evidence for the early evolution of microcystin synthesis. *Proc Natl Acad Sci U S A* **101**, 568-573.

Rapala J, Sivonen K, Lyra C, Niemela SI (1997) Variation of microcystins, cyanobacterial hepatotoxins, in *Anabaena* spp. as a function of growth stimuli. *Appl Environ Microbiol* **63**, 2206-2212.

Rinehart KL, Jr., Gloer JB, Hughes RG, Jr., *et al.* (1981) Didemnins: antiviral and antitumor depsipeptides from a caribbean tunicate. *Science* **212**, 933-935.

Rippka R, Deruelles J, Waterby JB, Herdmann M, Stanier RT (1979) Generic assignments, strain histories and properties of pure cultures of cyanobacteria. *J. Gen. Microbiol.* **111**, 1-61.

Rohrlack T, Christoffersen K, Hansen PE, et al. (2003) Isolation, characterization, and quantitative analysis of Microviridin J, a new *Microcystis* metabolite toxic to *Daphnia*. *J Chem Ecol* **29**, 1757-1770.

Rohrlack T, Christoffersen K, Kaebernick M, Neilan BA (2004) Cyanobacterial protease inhibitor microviridin J causes a lethal molting disruption in *Daphnia pulicaria*. *Appl Environ Microbiol* **70**, 5047-5050.

Rohrlack T, Dittmann E, Henning M, Börner T, Kohl JG (1999) Role of microcystins in poisoning and food ingestion inhibition of *Daphnia galeata* caused by the cyanobacterium *Microcystis* aeruginosa. *Appl Environ Microbiol* **65**, 737-739.

Rouhiainen L, Paulin L, Suomalainen S, et al. (2000) Genes encoding synthetases of cyclic depsipeptides, anabaenopeptilides, in *Anabaena* strain 90. *Mol Microbiol* **37**, 156-167.

Saitou N, Nei M (1987) The Neighbor-Joining Method - a New Method for Reconstructing Phylogenetic Trees. *Molecular Biology and Evolution* **4**, 406-425.

Sambrook J, Fritsch EF, Maniatis T (1989) *Molecular cloning. A laboratory manual* Cold Spring Harbour Lab Press.

Schmidt EW, Nelson JT, Rasko DA, et al. (2005) Patellamide A and C biosynthesis by a microcin-like pathway in Prochloron didemni, the cyanobacterial symbiont of *Lissoclinum patella*. *Proc Natl Acad Sci U S A* **102**, 7315-7320.

Schmidt HA, Strimmer K, Vingron M, von Haeseler A (2002a) TREE-PUZZLE: maximum likelihood phylogenetic analysis using quartets and parallel computing. *Bioinformatics* **18**, 502-504.

Schmidt HA, Strimmer K, Vingron M, von Haeseler A (2002b) TREE-PUZZLE: maximum likelihood phylogenetic analysis using quartets and parallel computing. *Bioinformatics* **18**, 502-504.

Schofield CJ, Baldwin JE, Byford MF, et al. (1997) Proteins of the penicillin biosynthesis pathway. *Curr Opin Struct Biol* **7**, 857-864.

Schopf JW (1993) Microfossils of the Early Archean Apex chert: new evidence of the antiquity of life. *Science* **260**, 640-646.

Schroeckh V, Scherlach K, Nutzmann HW, et al. (2009) Intimate bacterial-fungal interaction triggers biosynthesis of archetypal polyketides in *Aspergillus nidulans*. *Proc Natl Acad Sci U S A*.

Schwarzer D, Finking R, Marahiel MA (2003) Nonribosomal peptides: from genes to products. *Nat Prod Rep* **20**, 275-287.

Shin HJ, Murakami M, Matsuda H, Yamaguchi K (1996) Microviridins D-F, serine protease inhibitors from the cyanobacterium *Oscillatoria agardhii* (NIES-204). *Tetrahedron* **52**, 8159-8168.

Sinha Roy R, Kelleher NL, Milne JC, Walsh CT (1999) In vivo processing and antibiotic activity of microcin B17 analogs with varying ring content and altered bisheterocyclic sites. *Chem Biol* **6**, 305-318.

Stanier RY, Cohen-Bazire G (1977) Phototrophic prokaryotes: the cyanobacteria. *Annu Rev Microbiol* **31**, 225-274.

Stierle A, Strobel G, Stierle D (1993) Taxol and taxane production by Taxomyces andreanae, an endophytic fungus of Pacific yew. *Science* **260**, 214-216.

Strimmer K, von Haeseler A (1996) Quartet puzzling: A quartet maximum likelihood method for reconstructing tree topologies. *Mol. Biol. Evol.* **13**, 964-969.

Strimmer K, vonHaeseler A (1996) Quartet puzzling: A quartet maximum-likelihood method for reconstructing tree topologies. *Molecular Biology and Evolution* **13**, 964-969.

Sudek S, Haygood MG, Youssef DT, Schmidt EW (2006) Structure of trichamide, a cyclic peptide from the bloom-forming cyanobacterium *Trichodesmium erythraeum*, predicted from the genome sequence. *Appl Environ Microbiol* **72**, 4382-4387.

Takahashi I, Hayano D, Asayama M, et al. (1996) Restriction barrier composed of an extracellular nuclease and restriction endonuclease in the unicellular cyanobacterium *Microcystis* sp. *FEMS Microbiol Lett* **145**, 107-111.

Tan LT (2007) Bioactive natural products from marine cyanobacteria for drug discovery. *Phytochemistry* **68**, 954-979.

Thompson JD, Gibson TJ, Plewniak F, Jeanmougin F, Higgins DG (1997a) The CLUSTAL_X windows interface: flexible strategies for multiple sequence alignment aided by quality analysis tools. *Nucleic Acids Res* **25**, 4876-4882.

Thompson JD, Gibson TJ, Plewniak F, Jeanmougin F, Higgins DG (1997b) The ClustalX windows interface: flexible strategies for multiple sequence alignment aided by quality analysis tools. *Nucleic Acids Research* **25**, 4876-4882.

Tillett D, Dittmann E, Erhard M, et al. (2000) Structural organization of microcystin biosynthesis in *Microcystis* aeruginosa PCC7806: an integrated peptide-polyketide synthetase system. *Chem Biol* **7**, 753-764.

Trabi M, Craik DJ (2002) Circular proteins--no end in sight. *Trends Biochem Sci* **27**, 132-138.

Turner WB, Carter SB (1972) The chemistry and some biological effects of the cytochalasins. *Biochem J* **127**, 1P.

van der Meer JR, Polman J, Beerthuyzen MM, et al. (1993) Characterization of the Lactococcus lactis nisin A operon genes nisP, encoding a subtilisin-like serine protease involved in precursor processing, and nisR, encoding a regulatory protein involved in nisin biosynthesis. *J Bacteriol* **175**, 2578-2588.

Vergauwen B, De Vos D, Van Beeumen JJ (2006) Characterization of the bifunctional gamma-glutamate-cysteine ligase/glutathione synthetase (GshF) of Pasteurella multocida. *J Biol Chem* **281**, 4380-4394.

Vetter J (1998) Toxins of Amanita phalloides. *Toxicon* **36**, 13-24.

Vetting MW, LP SdC, Yu M, et al. (2005) Structure and functions of the GNAT superfamily of acetyltransferases. *Arch Biochem Biophys* **433**, 212-226.

Walsh V, Goodman J (1999) Cancer chemotherapy, biodiversity, public and private property: the case of the anti-cancer drug taxol. *Soc Sci Med* **49**, 1215-1225.

Weber T, Marahiel MA (2001) Exploring the domain structure of modular nonribosomal peptide synthetases. *Structure* **9**, R3-9.

Weeks B, Alcamo I (2007) *Microbes and Society*, 2. Edition edn. Jones & Bartlett Publishers.

Welker M, Brunke M, Preussel K, Lippert I, von Döhren H (2004) Diversity and distribution of *Microcystis* (Cyanobacteria) oligopeptide chemotypes from natural communities studied by single-colony mass spectrometry. *Microbiology* **150**, 1785-1796.

Welker M, von Döhren H (2006) Cyanobacterial peptides - nature's own combinatorial biosynthesis. *FEMS Microbiol Rev* **30**, 530-563.

Wenzel SC, Müller R (2005) Recent developments towards the heterologous expression of complex bacterial natural product biosynthetic pathways. *Curr Opin Biotechnol* **16**, 594-606.

Wenzel SC, Müller R (2007) Myxobacterial natural product assembly lines: fascinating examples of curious biochemistry. *Nat Prod Rep* **24**, 1211-1224.

Whitton BA, Potts M (2000)
. In: *The ecology of cyanobacteria: Their diversity in time and space* (eds. Whitton BA, Potts M), pp. Pp 1-11. Kluwer Academic, Boston.

Wilkinson B, Micklefield J (2007) Mining and engineering natural-product biosynthetic pathways. *Nat Chem Biol* **3**, 379-386.

Winz RA, Baldwin IT (2001) Molecular interactions between the specialist herbivore Manduca sexta (Lepidoptera, Sphingidae) and its natural host Nicotiana attenuata. IV. Insect-Induced ethylene reduces jasmonate-induced nicotine accumulation by regulating putrescine N-methyltransferase transcripts. *Plant Physiol* **125**, 2189-2202.

Zamble DB, McClure CP, Penner-Hahn JE, Walsh CT (2000) The McbB component of microcin B17 synthetase is a zinc metalloprotein. *Biochemistry* **39**, 16190-16199.

Zasloff M (1987) Magainins, a class of antimicrobial peptides from *Xenopus* skin: isolation, characterization of two active forms, and partial cDNA sequence of a precursor. *Proc Natl Acad Sci U S A* **84**, 5449-5453.

Zawadzke LE, Bugg TD, Walsh CT (1991) Existence of two D-alanine:D-alanine ligases in *Escherichia coli*: cloning and sequencing of the ddlA gene and purification and characterization of the DdlA and DdlB enzymes. *Biochemistry* **30**, 1673-1682.

Zerikly M, Challis GL (2009) Strategies for the discovery of new natural products by genome mining. *Chembiochem* **10**, 625-633.

Zhang CC, Laurent S, Sakr S, Peng L, Bedu S (2006) Heterocyst differentiation and pattern formation in cyanobacteria: a chorus of signals. *Mol Microbiol* **59**, 367-375.

Ziegler K, Diener A, Herpin C, et al. (1998) Molecular characterization of cyanophycin synthetase, the enzyme catalyzing the biosynthesis of the cyanobacterial reserve material multi-L-arginyl-poly-L-aspartate (cyanophycin). *Eur J Biochem* **254**, 154-159.

Ziemert N, Ishida K, Quillardet P, et al. (2008) Microcyclamide biosynthesis in two strains of *Microcystis aeruginosa*: from structure to genes and vice versa. *Appl Environ Microbiol* **74**, 1791-1797.

Zilliges Y (2007) *Molekulare Funktionsanalyse von Microcystin in Microcystis aeruginosa PCC 7806*, Doctoral Thesis. Humboldt-Universität zu Berlin.

Appendix

Amino acid alignment of MdnA homologues in *Microcystis* strains and field samples. Microviridin encoding sequences are shown in green and grey-green.

```
           **********  **********:*.*****.       * *:****  :
CBS        MAYPNDQQGKALPFFARPLSVSKEESSIKSPSPGHEIS---TRKYPSDWEEW-
MRC        MAYPNDQQGKALPFFARPLSVSKEESSIKSPSPDHEIS---TRKYPSDWEEW-
PCC9905    MAYPNDQQGKALPFFARPLSVSKEESSIKSPSPEHEIS---TRKYPSDWEEF-
Klon1      MAYPNDQQGKALPFFARPLSVSKEESSIKSPSPEPTYG--VTLKYPSDWEEF-
Klon4      MAYPNDQQGKALPFFARPLSVSKEESSIKSPSPEPTYG--VTLKYPSDWEEF-
Klon43     MAYPNDQQGKALPFFARPLSVSKEESSIKSPSPEPTYG--VTLKYPSDWEEF-
Klon46     MAYPNDQQGKALPFFARPLSVSKEESSIKSPSPEPTYG--VTLKYPSDWEEF-
Klon8      MAYPNDQQGKALPFFARPLSVSKEESSIKSPSPEPTYG--VTLKYPSDWEES-
Klon50     MAYPNDQQGKALPFFARPLSVSKEESSIKSPSPEPTYG--VTLKYPSDWEES-
Klon19     MAYPNDQQGKALPFFARPLSVSKEESSIKSPSPEPTYG--GTLKYPSDWEDY-
Klon25     MAYPNDQQGKALPFFARPLSVSKEESSIKSPSPEPTYG--GTLKYPSDWEDY-
Klon20     MAYPNDQQGKALPFFARPLSVSKEESSIKSPSPEPTFG--TTLKYPSDWEEY-
Klon23     MAYPNDQQGKALPFFARPLSVSKEESSIKSPSPEPTFG--TTLKYPSDWEEY-
MGL5tox    MAYPNDQQGKALPFFARPLSVSKEESSIKSPSPEPTFG--TTLKYPSDWEEY-
Nies298    MAYPNDQQGKALPFFARPLSVSKEESSIKSPSPEPTFG--TTLKYPSDWEEY-
Klon21     MAYPNDQQGKALPFFARPLSVSKEESSIKSPSPEPTTG--STPKYPSDWEDF-
Klon34     MAYPNDQQGKALPFFARPLSVSKEESSIKSPSPEPTYE--VTLKYPSDWEEF-
Klon31     MAYPNDQQGKALPFFARPLSVSKEEPSIKSPSPEREY---NVTLKYPSDWEEF-
Klon38     MAYPNDQQGKALPFFARPLSVSKEESSIKSPSPEREY---NVTLKYPSDWEEF-
Klon52     MAYPNDQQGKALPFFARPLSVSKEESSIKSPSPEREY---NVTLKYPSDWEEF-
Klon51     MAYPNDQQGKALPFFARPLSVSKEESSIKSPSPEREY---NVTLKYPSDWEEF-
Klon49     MAYPNDQQGKALPFFARPLSVSKEESSIKSPSPEREY---NVTLKYPSDWEEF-
Klon48     MAYPNDQQGKALPFFARPLSVSKEESSIKSPSPEREY---NVTLKYPSDWEEF-
Klon47     MAYPNDQQGKALPFFARPLSVSKEESSIKSPSPEREY---NVTLKYPSDWEEF-
Klon45     MAYPNDQQGKALPFFARPLSVSKEESSIKSPSPEREY---NVTLKYPSDWEEF-
Klon44     MAYPNDQQGKALPFFARPLSVSKEESSIKSPSPEREY---NVTLKYPSDWEEF-
Klon42     MAYPNDQQGKALPFFARPLSVSKEESSIKSPSPEREY---NVTLKYPSDWEEF-
Klon41     MAYPNDQQGKALPFFARPLSVSKEESSIKSPSPEREY---NVTLKYPSDWEEF-
Klon40     MAYPNDQQGKALPFFARPLSVSKEESSIKSPSPEREY---NVTLKYPSDWEEF-
Klon39     MAYPNDQQGKALPFFARPLSVSKEESSIKSPSPEREY---NVTLKYPSDWEEF-
Klon36     MAYPNDQQGKALPFFARPLSVSKEESSIKSPSPEREY---NVTLKYPSDWEEF-
Klon30     MAYPNDQQGKALPFFARPLSVSKEESSIKSPSPEREY---NVTLKYPSDWEEF-
Klon29     MAYPNDQQGKALPFFARPLSVSKEESSIKSPSPEREY---NVTLKYPSDWEEF-
Klon28     MAYPNDQQGKALPFFARPLSVSKEESSIKSPSPEREY---NVTLKYPSDWEEF-
Klon24     MAYPNDQQGKALPFFARPLSVSKEESSIKSPSPEREY---NVTLKYPSDWEEF-
Klon22     MAYPNDQQGKALPFFARPLSVSKEESSIKSPSPEREY---NVTLKYPSDWEEF-
Klon18     MAYPNDQQGKALPFFARPLSVSKEESSIKSPSPEREY---NVTLKYPSDWEEF-
Klon17     MAYPNDQQGKALPFFARPLSVSKEESSIKSPSPEREY---NVTLKYPSDWEEF-
Klon15     MAYPNDQQGKALPFFARPLSVSKEESSIKSPSPEREY---NVTLKYPSDWEEF-
Klon12     MAYPNDQQGKALPFFARPLSVSKEESSIKSPSPEREY---NVTLKYPSDWEEF-
Klon11     MAYPNDQQGKALPFFARPLSVSKEESSIKSPSPEREY---NVTLKYPSDWEEF-
Klon10     MAYPNDQQGKALPFFARPLSVSKEESSIKSPSPEREY---NVTLKYPSDWEEF-
Klon9      MAYPNDQQGKALPFFARPLSVSKEESSIKSPSPEREY---NVTLKYPSDWEEF-
Klon7      MAYPNDQQGKALPFFARPLSVSKEESSIKSPSPEREY---NVTLKYPSDWEEF-
Klon6      MAYPNDQQGKALPFFARPLSVSKEESSIKSPSPEREY---NVTLKYPSDWEEF-
Klon5      MAYPNDQQGKALPFFARPLSVSKEESSIKSPSPEREY---NVTLKYPSDWEEF-
Klon3      MAYPNDQQGKALPFFARPLSVSKEESSIKSPSPEREY---NVTLKYPSDWEEF-
Klon2      MAYPNDQQGKALPFFARPLSVSKEESSIKSPSPEREY---NVTLKYPSDWEEF-
Klon14     MAYPNDQQGKALPFFARPLSVSKEESSIKSPSPEREYGANVTLKYPSDWGEF-
Klon16     MAYPNDQQGKALPFFARPLSVSKEESSIKSPSPEREYGANVTLKYPSDWGEF-
Klon27     MAYPNDQQGKALPFFARPLSVSKEESSIKSPSPEREF---NVTLKFPSDWEEF-
KLON13     MAYPNDQQGKALPFFARPLSVSKEESSIKSPSPEREYS---TLKYPSDWEEF-
Klon26     MAYPNDQQGKALPFFARPLSVSKEESSIKSPSPEREYS---TLKYPSDWEEF-
Klon32     MAYPNDQQGKALPFFARPLSVSKEESSIKSPSPEREYS---TLKYPSDWEEF-
Klon33     MAYPNDQQGKALPFFARPLSVSKEESSIKSPSPEREYS---TLKYPSDWEEF-
Klon35     MAYPNDQQGKALPFFARPLSVSKEESSIKSPSPEREYS---TLKYPSDWEEF-
Klon37     MAYPNDQQGKALPFFARPLSVSKEESSIKSPSPEREYS---TLKYPSDWEEF-
PCC7005    MAYPNDQQGKAHPFFARPLSVSKEESSIKSPEEGRQ----TLKFPSDWEEF-
           1.......10........20........30........40........50...
```

Abbreviations

aa	amino acid
ABC	ATP-binding cassette
ACP	acyl carrier protein
Amp	Ampicilin
APS	ammonium persulphate
AT	Acyltransferase
ATP	adenosine triphosphate
bp	base pair(s)
BSA	bovine serum albumin
Cm	Chloramphenicol
DH	Dehydratase
DTT	1,4-dithiothreitole
DNA	deoxyribonucleic acid
dNTP	any desoxyribonucleotide
EDTA	ethylene diamine tetra-acetic acid
FITC	fluorescein isothiocyanate
HEPES	[4(2-hydroxyethyl)-1-piperazino]-ethanesulphonic acid
HGT	horizontal gene transfer
HPLC	high performance liquid chromatography
IFM	immunofluorescence microscopy

IPTG	isopropyl-thio-galactoside
JTT	Jones-Taylor-Thornton model for evolutionary rates
kb	kilo base pair(s)
kDa	kilo Dalton
KR	ketoacyl reductase
KS	ketoacyl synthase
MALDI	matrix-assisted laser desorption ionisation
NADH	nicotinamide adenine dinucleotide
NIES	National Institute for Environmental Studies
NJ	neighbor joining algorithm
NRPS	non-ribosomal peptide synthase
nt	nucleotide(s)
OD	optical density
ORF	open reading frame
PAGE	polyacrylamide gel electrophoresis
PBS	phosphate buffered saline
PCC	Pasteur Culture Collection
PCP	Peptidyl carrier protein
PCR	Polymerase chain reaction
PMSF	phenyl-methyl-sulphonyl-fluoride
RNA	ribonucleic acid
RP	ribosomal peptide

rpm	round per minute
RT	room temperature
SDS	sodium dodecyl sulphate
TBE	tris-borate-EDTA buffer
TEMED	N',N',N',N'-tetramethyl-ethylene-diamine
TOF	time of flight
Tris-HCl	tris-(hydroxymethyl)-aminomethane-hydrochloride
UV	ultraviolet light
UWOCC	University of Wisconsin at Oshkosh Culture Collection
WT	wild type
X-Gal	5-bromo-4-chloro-3-indolyl-β-D-galactopyranoside

Eigenständigkeitserklärung

Gemäß §6 der Promotionsordnung der Mathematisch-Naturwissenschaftlichen Fakultät I der Humboldt-Universität zu Berlin

Hiermit erkläre ich, dass ich die vorliegende Arbeit selbständig angefertigt und ohne fremde Hilfe verfasst habe, keine außer den angebenen Hilfsmitteln und Quellen dazu verwendet habe und die den benutzten Werken wörtlich oder inhaltlich entnommenen Stellen als solche kenntlich gemacht habe.

Nadine Ziemert

Acknowledgement

Ein großer Teil meines Dankes gilt meiner Betreuerin Frau Prof. Elke Dittmann. Ihr Interesse und ihre Begeisterung an der Wissenschaft sind ansteckend.

Vielen Dank an die Mitarbeiter der Molekularen Ökologie, Genetik und Molekularen Genetik für die nette Arbeitsatmosphäre. Vor allem Arthur Guljamow und Annika Weiz möchte ich für die tolle Zusammenarbeit und die nie langweilig werdenden Stunden im Labor danken. Ein großes Danke an Jan - Christoph Kehr für die Hilfe, vor allem auch bei Computerproblemen. Danke auch an die praktische Unterstützung von Jana Müller, Katrin Hinrichs und Ramona Günther im Labor. Für die große Hilfe in allen organisatorischen Fragen möchte ich Petra Dreier danken. Die Arbeit mit Euch allen hat Spaß gemacht.

Many thanks to Dr. Keishi Ishida and Prof. Christian Hertweck for the great collaboration and the time and energy they have dedicated to all the chemistry and structure elucidations, which form an important constituent of this thesis.

I also like to thank Dr. Anton Liaimer for the pleasant collaboration in the microviridin project.

Danke an meine Freunde, dass ihr da seid. Danke an meine große und bunte Familie, dass ich mich nie alleine fühle.

Danke Oli für alles.

Die VDM Verlagsservicegesellschaft sucht für wissenschaftliche Verlage abgeschlossene und herausragende

Dissertationen, Habilitationen, Diplomarbeiten, Master Theses, Magisterarbeiten usw.

für die kostenlose Publikation als Fachbuch.

Sie verfügen über eine Arbeit, die hohen inhaltlichen und formalen Ansprüchen genügt, und haben Interesse an einer honorarvergüteten Publikation?

Dann senden Sie bitte erste Informationen über sich und Ihre Arbeit per Email an *info@vdm-vsg.de*.

Sie erhalten kurzfristig unser Feedback!

VDM Verlagsservicegesellschaft mbH
Dudweiler Landstr. 99   Telefon +49 681 3720 174
D - 66123 Saarbrücken   Fax   +49 681 3720 1749
www.vdm-vsg.de

Die VDM Verlagsservicegesellschaft mbH vertritt

Printed by Books on Demand GmbH, Norderstedt / Germany